Food and Packaging Interactions II

Food and Packaging Interactions II

Sara J. Risch, EDITOR
Golden Valley Microwave Foods, Inc.

Joseph H. Hotchkiss, EDITOR
Cornell University

Developed from a symposium sponsored
by the Division of Agricultural and Food Chemistry
at the 200th National Meeting
of the American Chemical Society,
Washington, D.C.,
August 25–31, 1990

American Chemical Society, Washington, DC 1991

Library of Congress Cataloging-in-Publication Data

Food and packaging interactions II / Sara J. Risch, editor, Joseph H. Hotchkiss, editor.

    p.   cm.—(ACS symposium series; 473)

    "Developed from a symposium sponsored by the Division of Agricultural and Food Chemistry at the 200th National Meeting of the American Chemical Society, Washington, D.C., August 25–31, 1990."

    Includes bibliographical references and indexes.

    ISBN 0–8412–2122–7

    1. Food—Packaging—Congresses. 2. Food—Analysis—Congresses.

    I. Risch, Sara J., 1958– . II. Hotchkiss, Joseph H. III. American Chemical Society. Division of Agricultural and Food Chemistry. IV. American Chemical Society. Meeting (200th: 1990: Washington, D.C.). V. Title: Food and packaging interactions 2. VI. Title: Food and packaging interactions two. VII. Series.

TP374.F643   1991
664′.092—dc20   91–31831
                                  CIP

The paper used in this publication meets the minimum requirements of American National Standard for Information Sciences—Permanence of Paper for Printed Library Materials, ANSI Z39.48–1984. ∞

ACS Symposium Series

M. Joan Comstock, *Series Editor*

1991 ACS Books Advisory Board

Foreword

THE ACS SYMPOSIUM SERIES was founded in 1974 to provide a medium for publishing symposia quickly in book form. The format of the Series parallels that of the continuing ADVANCES IN CHEMISTRY SERIES except that, in order to save time, the papers are not typeset, but are reproduced as they are submitted by the authors in camera-ready form. Papers are reviewed under the supervision of the editors with the assistance of the Advisory Board and are selected to maintain the integrity of the symposia. Both reviews and reports of research are acceptable, because symposia may embrace both types of presentation. However, verbatim reproductions of previously published papers are not accepted.

Contents

INDEXES

Preface

Vast development of new packaging materials with a broad range of uses has occurred since 1987. In addition to protecting food products, packages now often serve as cooking containers. The combination of new materials and new uses for packages has resulted in previously unrecognized interactions between foods and the packages containing them.

Food–packaging interactions can occur in a number of different ways. Components of the package can move into the food, a phenomenon that is called migration. Food components may also be absorbed by the package, and when the package absorbs flavor compounds the process is called scalping. The package may also influence or induce reactions in foods. An understanding of the permeability of packaging materials is important when selecting materials to provide adequate moisture, oxygen, and aroma barriers.

Susceptor packaging, which contains specialized material that heats in a microwave oven to temperatures that will help products to brown and crisp, has received considerable attention in the past few years. Although there has been no health concern about this type of packaging, microwave suseptors have been researched extensively to confirm their safety and to provide information for the development of appropriate regulations. A significant portion of the symposium upon which this book is based covered the research that deals with susceptor packaging.

This book probably provides the most comprehensive and current information in the field of food–packaging interactions all in one publication. This volume can serve as a resource for food product development scientists and packaging engineers who need to understand what can happen to foods in different types of packages. Regulatory agencies can also use this book to keep pace with new developments and to maintain appropriate regulations.

We thank all of the authors who contributed their time and energy to make this book possible.

SARA J. RISCH
Golden Valley Microwave Foods, Inc.
6866 Washington Avenue South
Eden Prairie, MN 55344

JOSEPH H. HOTCHKISS
Institute of Food Science, Stocking Hall
Cornell University
Ithaca, NY 14853

June 28, 1991

Introduction

Packaging, now an integral part of the food chain, has gone from simply being a convenient container to hold and carry food to highly specialized systems serving many purposes. Many packaging materials still are little more than carrying containers that serve as dust covers to protect food from outside contamination. The demand for convenience and quality in food products has been accompanied by the development of packaging materials to help extend the shelf life of products; to provide barriers to moisture, aroma, and modified atmospheres; and to serve as cooking containers in both conventional and microwave ovens. Plastics, which can be one specific material (homopolymers) or a combination of materials (copolymers) and which can be laminated or coextruded, make up a large percentage of new packaging materials. Other packaging materials include paper or foil laminated to plastic film and tied together by adhesive layers. Various combinations of these materials can be used to meet the requirements of specific food products.

Glass, which provides barrier properties, allows the preservation of food, and can also serve as a cooking container, was one of the first food containers. For packaging purposes, glass is considered virtually inert, that is, it does not interact with or change the characteristics of the food with which it comes into contact. This is not always the case with newer packaging materials such as plastic, paper, and paperboard. These new materials can interact with food products in a variety of ways.

Permeability of these materials to various food components is one area of interest. Plastic overwraps or laminations are often used either to keep moisture in a product or to prevent the product from picking up moisture from the surrounding atmosphere. In addition to being a moisture barrier, the package may also serve as an aroma barrier to help retain flavor in foods and to prevent the absorption of undesirable flavors or aromas. Although package permeability may result in flavor loss, an area that has received considerable attention is the absorption of flavors or specific flavor components by packaging materials. Flavor compounds tend to have low molecular weights and, as a result, not only pass more readily through plastic films but are also soluble in the polymers from which the films are made. This loss of flavor to the packaging is known as scalping. Individual compounds that make up a flavor can be selectively sorbed from a food product, with different compounds absorbed by the package to different extents. For example, 5% of compound X may be absorbed while 30% of compound Y is absorbed. The result is both an unbalanced flavor in the product and an overall decrease in flavor intensity.

A number of aspects of flavor scalping can be investigated. One area of interest is the relationship between a given type of film and a wide variety of flavor compounds to determine how the flavor of a product might change if put in contact with that particular polymeric structure. Another approach is to look at a particular class of compounds, for example, aldehydes or acids, and to determine how that particular class interacts with a variety of different films. Molecular weight and boiling point are properties that can influence the rate or extent to which flavor compounds are adsorbed by packages.

Studies of scalping will help to determine the types of plastic films best suited for particular food products. Although some films may be adequate moisture or gas barriers for a product, that benefit can be offset by scalping. Flavors may be reformulated to compensate for selective loss of flavor compounds into the package, but the rate and extent of flavor loss, and the point at which an equilibrium is reached between the amount of a compound in the food and in the package must be known for reformulation to be successful.

Components of the packaging material may also migrate from the package into the food. This is of concern for both the organoleptic quality and the safety of the food. When polymers for plastic films and packages are produced, complete polymerization of the monomers or starting units, although technically possible, it is not economically feasible. Incomplete polymerization results in residual monomers or oligomers in the finished materials. For example, residual styrene may be present in polystyrene, and the cyclic trimer may be present in poly(ethylene terephthalate).

In addition to the monomers and oligomers, many materials may contain small percentages of other low-molecular-weight compounds that are often referred to as processing aids. A number of different types of compounds, including phthalates, are added as plasticizers. These plasticizers promote flexibility in the film and make the film easier to process. Some polymerization reactions require initiators or catalysts. Other compounds may be added to plastics to improve the strength, color, or other properties of the package.

All of these low-molecular-weight materials are capable of migrating into the food with which they come into contact. This is an active area of study to ensure the wholesomeness of food products in these various packages. In the United States, extensive regulations that cover most existing packages and their intended use are outlined in the *Code of Federal Regulations*. As new materials are developed and new applications are found for existing materials, further research must be performed to confirm the safety of the food packaging.

As was previously mentioned, the migration of low-molecular-weight compounds can also impact the organoleptic quality of the food. Some compounds that migrate may not pose a health risk, but they may create an off flavor in the food. Therefore, packages must be screened not only from a safety standpoint but also from the standpoint of the sensory quality of the food.

One development that has received considerable attention during the past few years is the microwave susceptor, which is a package developed to help meet the demands of convenience and quality in microwaveable food products. A susceptor is typically a piece of polyester film that has been metallized with aluminum. This material will interact with microwave energy to heat rapidly to temperatures in the 400 ° F range. Susceptors promote browning and crisping of products in the microwave oven and help popcorn to pop more efficiently. The susceptor is laminated either to paperboard or between two layers of paper. All of the materials used in susceptor packaging meet existing regulations; however, the existing regulations were designed for usage conditions where much lower temperatures were anticipated. Although there have been no health or safety concerns, susceptor packages have been studied extensively to determine if new regulations are necessary and, if so, what regulations would be appropriate.

Materials that can withstand the higher temperatures encountered in a conventional oven and still perform in a microwave oven are also being studied. These materials must also be evaluated to ensure that they do not create off flavors in foods and that package components do not migrate into the food.

These are some of the more significant areas being studied in the field of food–package interactions. As new packaging materials are developed and new uses are found for existing materials, research into the safety and effectiveness of food packaging will continue.

SARA J. RISCH
Golden Valley Microwave Foods, Inc.
6866 Washington Avenue South
Eden Prairie, MN 55344

June 28, 1991

Chapter 1

Analysis of Volatiles Produced in Foods and Packages during Microwave Cooking

Sara J. Risch[1], Kurt Heikkila[2], and Rodney Williams[2]

[1]Golden Valley Microwave Foods, Inc., 6866 Washington Avenue South, Eden Prairie, MN 55344
[2]Aspen Research Corporation, 436 West County Road D, New Brighton, MN 55112

Specific methodology has been developed to qualitatively and quantitatively evaluate volatile compounds produced during microwave cooking. A closed system was designed to cryogenically trap volatile compounds produced during microwave heating outside of the microwave oven. Volatiles generated by foods and packaging materials can be evaluated to determine what is generated and where it goes. Claims for a patent covering the method and apparatus have been allowed. The application of this method specifically to microwave popcorn bags will be discussed.

The food and packaging industries have been working on new technologies to improve the quality and convenience of microwavable and dual-ovenable products. One significant challenge in this area has been to find a way to help foods brown and crisp in the microwave. The development that has most successfully met this challenge is a structure referred to as a susceptor. The term susceptor encompasses a number of different structures that are used as microwave cooking or heating containers and reach temperatures of approximately 350-450°F during use.(1) The vast majority of susceptors used are thin film susceptors composed of a plastic film, typically polyester (polyethylene terephthalate) that is coated with a thin layer of metal. The most commonly used metal is aluminum, however, stainless steel is also used. The susceptor is incorporated into a package in predominantly two distinct constructions. In one patented construction (2) the thin film susceptor is laminated between two layers of paper to prevent direct contact

0097–6156/91/0473–0001$06.00/0

of the susceptor with the food product. The predominant use of this construction is for microwave popcorn bags, although it is also being used for other applications including some varieties of microwave bread, waffles and french fries. Microwave popcorn bags represent approximately 75% of all susceptors used (*3*). In the other principle construction, the susceptor is laminated directly to paperboard and is in direct contact with the food. It should be noted these two constructions do not encompass all of the susceptors used today, however, they represent over 95% of the market. There are also other proposed constructions which have been developed but are not currently in commercial use.

Considerable attention has been focused on the higher temperatures that susceptors reach during use. The main issues have been the absence of regulations for materials that reach temperatures greater than 300°F during use and the lack of data about the behavior of these materials at the higher temperatures. The data is necessary to write appropriate regulations and confirm the safety of the packages. A method was necessary to develop data to qualitatively and quantitatively study the components of the packaging materials.

MATERIALS AND METHODS

A test apparatus was designed to contain a microwave package during cooking. The apparatus is a closed purge and trap type of system that allows volatile compounds generated by the package and/or food to be trapped outside of the oven. A diagram of the test cell is shown in Figure 1. A purge line runs in one orifice at the end of the cell. This line runs to the other end of the cell. A vent line goes from the other orifice to a cryo-trap outside the oven. This is representative of what actually happens during microwave cooking. Steam and volatile compounds are driven into the cavity of the microwave oven and are vented out of the oven with a fan that circulates air through the oven.

For popcorn, the bag is placed in the cell with the end that vents at the end of the test cell where the vent line is located. The purge line is at the other end of popcorn bag to gently sweep out volatiles evolved from the package, but not sweep out any volatiles that would remain in the bag during normal popping. The purge flow can be varied depending on the product to be tested. Experimental work showed that a flow of 1 L/min of nitrogen

Reactor System

Developed for Popcorn/Oil/Bag Volatile Analyses

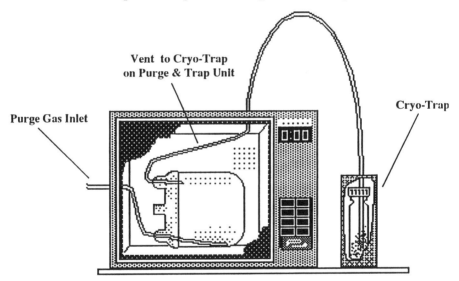

Figure 1. Test cell developed to contain a package and/or food product during microwave cooking.

was most effective for popcorn. The cryo-trap consists of a vessel packed with glass wool which facilitates trapping of the compounds which are present in an excess of steam. Following the purging of the test cell during microwave heating, the trap is transferred to a traditional purge and trap system (O.I. Corporation) attached to a Hewlett Packard model 5890 gas chromatograph (GC) coupled to a Hewlett Packard model 5970 mass selective detector (MS).

The volatile compounds generated by the susceptor containing portion of a microwave popcorn bag were initially identified using headspace gas chromatography mass spectrometry (GC/MS). The procedure was adapted from ASTM procedure F1308-90. Approximately 1 g of packaging material was sealed in a headspace vial. The vial was heated in a nominal 700 watt microwave oven on high power for 3.5 minutes. A 250 g water load in a glass beaker was placed in the oven with the headspace vial.

Temperature profiles were run during microwave heating using a Luxtron Fluoroptic Thermometry System. Temperatures of approximately 232°C (450°F) were reached. This represents the highest temperatures normally attained during preparation of microwave popcorn.

The headspace vial was cooled in liquid nitrogen before transferring it to a heated purge and trap system (O.I. Corporation). This deviation from the ASTM static headspace method was added to achieve greater sensitivity. The vial was purged for 7 minutes while being held at 90°C. The trap was desorbed directly onto the GC column in the GC/MS system described above. The volatile compounds were identified and quantified. The identity was confirmed with reference standards.

After identifying the volatiles generated by the package, a series of experiments was run using the test cell to determine the volatiles generated by the individual components used in microwave popcorn and the complete product. Oil, popcorn and oil, the package and popcorn and oil in a microwave popcorn bag were each heated in the test cell in a microwave oven. The volatiles generated by each set of components were trapped and analyzed as described earlier for the test cell. This allowed for quantitation of volatiles generated by the package and food during the microwave cooking process. For the oil alone and popcorn and oil, a susceptor was placed on the outside of the test cell to generate equivalent heat to that attained during popping in the bag.

Temperature profiles of the popcorn bag being heated in the test cell were run using a Luxtron (Mountain View, California) with an MIH probe. A representative profile is shown in Figure 2. The profile is similar to that generated by a popcorn bag being heated in the same microwave oven but

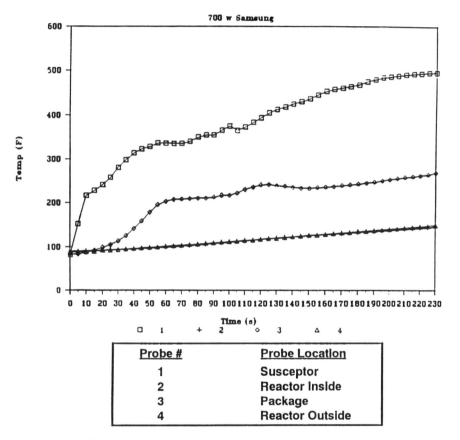

Figure 2. Temperature profile of microwave popcorn bag and test cell being heated in a microwave oven.

not in the test cell, although the temperatures attained are slightly higher in the test cell. This indicates that the testing will represent the worst case for cooking of microwave popcorn.

Recoveries were determined by using target compounds (key compounds identified as being generated by the package), internal standards and surrogate materials. The compounds were spiked into the test cell or into the popcorn and oil. The recovery experiments were based on EPA methodology detailed in USEPA document SW-846, Method 8240 volatile analysis.

The final procedure was to determine if any of the volatiles could be recovered from popcorn popped in a microwave popcorn bag. One target compound, benzene, was chosen for the recovery study. The deuterated analog, D_6benzene was used as a surrogate standard to prove system recovery. The test cell used in the microwave oven was placed in a heat jacket and maintained at 105°C. A 10 ul aliquot of a solution containing 11 ppb D_6benzene was put into the sealed test cell. The cell was purged for 5 minutes with nitrogen at 1 L/min into a trap packed with glass wool and held in liquid nitrogen. After purging, the trap was transferred to the purge and trap on the GC/MS.

Recovery from popcorn was determined by transferring approximately 1 L of popped corn that was prepared in a microwave popcorn bag into the test cell. The cell was spiked with 10 ul of 11 ppb D_6benzene solution, purged and analyzed as described above.

RESULTS AND DISCUSSION

The screening test on the susceptor portion of microwave popcorn bags showed that volatile compounds are generated during microwave heating, however, the amounts generated are quite small. A list of the compounds identified is contianed in Table 1. Figure 3 shows a representative total ion chromatogram. From this list, compounds were selected for the serial experiments in the test cell. The compounds selected were those that were either present in the largest amounts or of greatest toxicological concern. Input from FDA was obtained in determining the appropriate compounds. These compounds were: benzene, toluene, n-butyl ether, 2-furancarboxaldehyde (furfural), styrene, 2-ethyl-l-hexanol and dodecane.

Table 1
Compounds Identified In Popcorn After Microwave Heating

Peak #	Compound
1	2-Methyl-2-Propanol
2	N-Methyl-Formamide
3	2-Methyl-Furan
4	2-Butanone
5	Benzene
6	2-Methyl-2-Propenal
7	2-Butenal
8	2,3-Dihydro-1,4-Dioxane
9	2,3-Pentanedione
10	Phenol
11	1,4-Dioxane
12	2-Cyclopenten-1-one
13	3-Methyl-Furan
14	2-Methyl-1-Propene
15	2-Furancarboxaldehyde
16	Bicyclo[4.2.0]octa-1,3,5-triene
17	1-Ethenyl-2-Methyl-Benzene
18	2,6-Dimethyl-Nonane
19	(1-Methylethenyl)-Benzene
20	Benzaldehyde
21	1-Methyl-2-(2-Propenyl)-Benzene
22	Undecane
23	6Mmethyl-Dodecane
24	1-Phenyl-1,2-Propanedione
25	Dodecane
26	6-Methyl-Dodecane
27	Tridecane
28	2,7,10-Trimethyl-Dodecane
29	Tetradecane

The results from the determination of the volatiles generated by the oil, corn and oil and complete product in the package are summarized in Table 2. These results were obtained using the test cell and represent the compounds trapped outside of the microwave oven. It should be noted that both the corn and oil generate volatile compounds during microwave heating. The most notable of these is furfural. This is a compound that is generated as a heat degradation product of cellulose and was questioned as a potential concern from susceptors. Furfural is a commonly used flavor compound and is generated during browning of food products in the reactions between amino acids and sugars. As can be seen, popcorn in oil generated the equivalent of 26.3 ug/in^2 furfural, which is about the same as the 16.2 ug/in^2 generated by the complete package. Furfural is naturally produced by the product itself and should not raise any issues concerning the safety of the package. Furfural was not included in any further analyses.

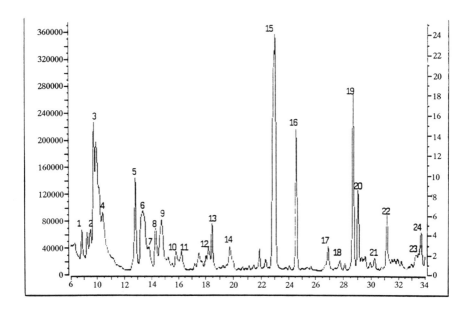

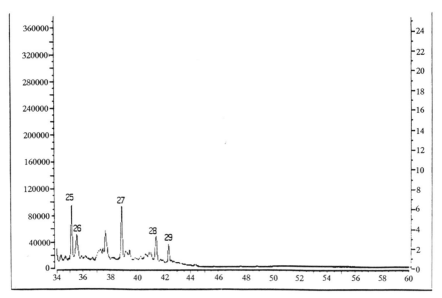

Figure 3.   Representative total ion chromatogram for volatile compounds generated by popcorn bags during microwave heating.

Table 2
Selected Volatile Constituents Evolved From Oil,
Oil And Corn, And Complete Package

Analyte	Product In Package	Oil	Oil & Corn	STD 0330
Bromochloramethane*	0.16	0.16	0.16	0.16
1,4-Difluorobenzene*	0.16	0.16	0.16	0.16
Benzene	0.0057	0.00032	nd	0.18
Toluene	0.019	0.013	0.011	0.16
Chlorobenzene-o5*	0.16	0.16	0.16	0.16
n-Butyl Ether	0.0187	0.00061	nd	0.11
2-Furancarboxaldehyde	16.23	nd	26.33	30.08
Styrene	0.008	nd	0.00070	0.20
Benzaldehyde	nd	nd	nd	4.76
2-Ethyl-1-Hexanol	0.45	nd	0.00026	1.61
Dodecane	0.27	0.0020	0.0076	1.61

* Denotes Internal Standard
All concentrations in ug/in2

These results indicate that the volatiles generated during microwave heating of popcorn vent out of the bag and can be trapped outside of the oven for quantification. To help explain the phenomenon that the volatiles are driven out of the bag and do not transfer to the popcorn, an experiment was run to determine the amount of steam generated during popping of the corn. An average of 6 g of weight was lost from a 50 g bag during popping. Assuming that the weight lost is virtually all steam and using the gas law for conversion, an average of 10158 cm^3 of steam was generated. The average volume of the bag is 1460 cm^3. This translates into approximately 7 bag volumes of steam generated during popping. With this large volume of steam, the volatiles are steam distilled out of the bag.

The final experiment confirmed that benzene generated by the package does not transfer into the popcorn. The results of seven replicate trials are listed in Table 3. The system recovery for D_6benzene averaged 40% when only the standard was added. Similar recoveries were obtained when popcorn was added to the system. With a calculated minimum detection limit of 0.0012 ug/in^2, no benzene was detected in the popcorn.

The procedure and apparatus developed were effectively used to both qualitatively and quantitatively look at volatile compounds generated during microwave heating. While it was used for popcorn, it could have applications with all microwave packaging material and foods. It could also be used to investigate volatiles generated during conventional heating.

Table 3
Determination of Benzene and Benzene-d6

Run#	Benzene ug/Kg	Benzene ug/sq.inch	Benzene-d6 ug/Kg	Benzene-d6 Recovery %
Spiked Popcorn 1	ND	ND	6.0	40
Spiked Popcorn 2	ND	ND	6.1	40
Spiked Popcorn 3	ND	ND	6.1	41
Spiked Popcorn 4	ND	ND	5.1	38
Spiked Popcorn 5	ND	ND	5.3	35
Spiked Popcorn 6	ND	ND	7.7	51
Spiked Popcorn 7	ND	ND	5.6	36
Spiked Blank 1	ND	ND	7.7	52
Spiked Blank 2	ND	ND	7.9	53
MDL	0.24	0.0012		

	Average Benzene-d6 Recovery %	Std. Deviation of Benzene-d6 Recoveries %
Spiked Popcorn - 7 Runs	40	5.3

Literature Cited

1. Kashtock, M.E.; Wurts, C.B.; Hamlin, R.N. 1990. J. Packaging Technology, Vol. 4, No. 2: 14-19.
2. Watkins, J.D.; Andreas, D.W.; Cox, D.H. 1988. U.S. Patent 4,735,513.
3. Larson, Melissa, 1989. Packaging Vol. 34, No. 16: 51-52.

RECEIVED June 21, 1991

Chapter 2

Migration into Food during Microwave and Conventional Oven Heating

Sue M. Jickells[1], John W. Gramshaw[2], John Gilbert[1], and Laurence Castle[1]

[1]Ministry of Agriculture, Fisheries, and Food, Food Science Laboratory, Colney Lane, Norwich NR4 7UQ, United Kingdom
[2]Procter Department of Food Science, University of Leeds, Leeds LS2 9JT, United Kingdom

Novel applications of packaging for microwave oven use and introduction of new innovative materials make it evermore important that migration testing is systematically and rigorously carried out by regulatory authorities. A testing regime for packaging and cookware articles is described, which is seen as complementary to research-oriented investigations to assess the identity and migration levels of trace organic species present in these materials. A number of examples are used to illustrate the range of species that may be present in packaging and other food contact materials and the approaches adopted for their analysis in foods under actual conditions of use.

Future regulatory control of food contact materials within the European Community, will increasingly depend on the use of positive lists of permitted substances supported by migration testing (1,2). The use of a particular monomer or additive in a plastic will be controlled, and the amount that migrates will also be regulated. For the purpose of operating these statutory controls, most migration testing will employ simulants rather than real foods as this simplifies analysis and permits large numbers of samples to be analysed on a routine basis.

In addition to positive lists and specific migration limits, general provisions were introduced under a 1990 European Community Directive (3) controlling the suitability of materials and articles on the basis of overall migration limits. The Directive states that plastics materials and articles shall not transfer their constituents to foodstuffs in quantities exceeding 10 mg/dm^2 of surface area of material or article. For certain situations this limit is set at 60 mg/kg of foodstuff. As for specific migration testing, compliance with these overall limits will be assessed using food simulants.

0097–6156/91/0473–0011$06.00/0
© 1991 American Chemical Society

The conditions to be used for testing are specified in the Directive. The choice of appropriate food simulant and the time and temperature of exposure of the plastic to the simulant, depends on the expected use of the plastic. The test conditions are summarised in Table 1 and are applicable to a maximum temperature of 121°C during actual use.

Table 1. Contact conditions to be used for testing migration from plastics materials according to contact conditions during actual use (4)

Contact conditions during actual use	Test conditions
1. Duration of contact: more than 24 h	
T <5°C	10 days at 5°C
T >5°C but <20°C (labelling mandatory)	10 days at 20°C
T >5°C but <40°C	10 days at 40°C
2. Duration of contact: between 2 and 24 h	
T <5°C	24 h at 5°C
T >5°C but <40°C	24 h at 40°C
T >40°C	in accordance with national legislation
3. Duration of contact: less than 2 h	
T <5°C	2 h at 5°C
T >5°C but <40°C	2 h at 40°C
T >40°C but <70°C	2 h at 70°C
T >70°C but<100°C	1 h at 100°C
T>100°C but<121°C	30 min at 121°C
T>121°C	in accordance with national legislation

For the intended use of food contact plastics above 121°C testing is "in accordance with national legislation", which in the UK is under the general provisions of the 1987 Materials and Articles in contact with Food Regulations (5). In order to extend the migration testing Scheme in Table 1 to encompass the high temperature use of plastics, it became apparent that it was necessary to establish information on the identity of commercial materials intended for these applications and to measure the actual temperatures achieved at the food/plastic interface during heating of foods. This information could then be used for proposing a testing scheme with simulants and the scheme proposed could in turn be validated by comparing test results with migration levels into real foods. For microwave applications it was also apparent that the effect (if any) of microwave energy itself on migration would need to be investigated as this would determine whether migration testing would or would not require the use of microwave heating.

Identity of plastic materials employed in the UK for microwave and conventional oven use

The use of ovenable plastics falls broadly into two categories:

A. Prepacked items, where the article is filled with food by the food manufacturer, for subsequent reheating or cooking by the consumer.

B. Cookware articles and films, where there is no food contact at the point of sale. The consumer may put such cookware to a variety of uses in food preparation.

Items in category A for microwave use include polyethylene terephthalate (PET), polypropylene (PP) laminates with a barrier layer of ethylene/vinyl alcohol copolymer (EVOH) or poly(vinylidene chloride) (PVdC), polyethylene (PE) and susceptor materials. Susceptors are designed to interact with microwave radiation and so crisp and brown foods by localised heating. This is achieved by means of a thin layer of aluminum vacuum deposited onto polyester film. The metallized film is laminated to a paper or paperboard substrate. The polyester or the paper can be the food contact surface. Plastics for conventional oven use include PET and various laminated paperboard materials, some again with PET as the food contact layer.

Cookware for home use in conventional ovens (category B) includes thermoset polyester (TPES), PET, poly(methylpentene) (TPX) and poly(etherimide). These materials are also marketed for microwave use and the range of materials is extended further by PE, PP, polycarbonate and polysulphone. Films on retail sale include PE and plasticised materials prepared from either vinylidene chloride/vinyl chloride (VdC/VC) copolymer or PVC. PVC and PE films bear 'on-the-pack' labelling which instructs that the films should not be used in conventional ovens nor should they be used for wrapping foods or lining dishes during microwave cooking but may be used for covering containers or reheating meals on a plate. Also on retail sale is a susceptor-type material with PET as the food contact layer laminated to a flexible paper backing.

Temperatures achieved at the food-plastic interface during microwave and conventional oven cooking

Temperatures were measured using a fiber optic thermometer (Asea Model 1010). A large number of pre-packaged foodstuffs and recipes taken from cookbooks were cooked or reheated according to instructions and some specimen results are presented in Table 2. For foods which were largely aqueous in nature, temperatures did not generally exceed 100°C. For some foodstuffs, however, particularly those which had a high fat or sugar content, 121°C (the maximum EC test temperature) was exceeded.

Table 2. Temperatures recorded at the food/container interface
during a variety of cooking situations in microwave and
conventional ovens

Food Type	Time (min)	Max Temp (^{o}C)
A. Microwave cooking of home-prepared foodstuffs [a]		
Lasagne sauce	10	101
Spinach	4	100
Fruit sauce	5	101
Roast Chicken	37	106
Butterscotch	5	123
Meringue	20	150
Peanut brittle	24	150
Bacon	3-4	200
B. Microwave cooking of pre-packed foodstuffs [a]		
Fish-in-sauce	4+4 [b]	115
Fish fillets	3+3 [b]	102
Pasta and sauce	7	121
Chilli-con-carne	5	99
Chocolate pudding	5	101
C. Conventional oven cooking of pre-packaged foodstuffs [c]		
Pasta and sauce	25	99
Chicken rissoto	35	92
Lasagne	40	99
Garlic bread	15	194

(a) Time at 600W power
(b) + standing time
(c) Electric oven. Pasta 175°C, chicken and lasagne 205°C, bread
     250°C.

Effect of microwave energy on migration from food contact materials

Migration testing at elevated temperatures is normally conducted
in a calibrated laboratory oven. This is adequate for materials
which will be cooked or processed using heat from a conventional
oven. However, it is questionable whether conventional heating
should be used in determining migration from materials intended for
microwave oven use. If the only influence of microwave energy on
migration is thermal then conventional heating is acceptable but if
there is any non-thermal influence, testing will need to be carried
out in a microwave oven. A series of experiments was therefore
conducted to investigate non-thermal effects of microwave energy
on migration. Five materials were selected for study, namely:
PET, PP, TPES, TPX and PVdC/PVC. Experimental techniques were

designed to measure any influence of microwave energy on migration, over and above that produced by normal heating processes. Two experimental approaches were taken:

1. Pre-treatment of the plastic in a microwave oven, followed by migration testing using simulant in a conventional laboratory oven. Comparison of results with those for an unexposed (non-microwaved) plastic tested for migration under identical conditions.

2. 'Real-time' migration with the material in contact with simulant and exposed to microwave radiation, holding the simulant at a constant temperature. Comparison of results for material exposed to simulant under precisely the same time and temperature conditions but using conventional rather than microwave heating.

Olive oil was used as the simulant for Approach 1 and iso-octane was used as the simulant for Approach 2. To prevent damage, the oven for Approach 1 was modified by the installation of cooling coils (40 ml capacity), through which $18^{\circ}C$ water flowed at 17 ml/s. The microwave pre-treatment was for 12 hours at a duty cycle of 38%, giving an equivalent of 4.5 h at 575 W measured according to IEC 705 (16). For Approach 2, the plastic and simulant were held in a jacketed vessel and microwaved at full power (600 W). Chilled water was circulated through the vessel at 17 ml/s to counterbalance the heat input from the radiation. The test plastic was thus held in the simulant at a constant temperature for 30 min. Experiments to compare migration using conventional heating were conducted in a water bath using the same time and temperature. Migration of a specific compound was measured for each material (see Table 3). An outline of the methods used for the determination of individual migrants is also shown in Table 3.

The results of migration from microwave treated and non-treated materials are shown in Table 4. Migration from non-treated material has been normalized as 100% and the migration from the microwave-treated material given relative to this. Within the analytical precision of each method, there were no statistically significant differences ($p < 0.05$) in migration from microwave-treated and non-treated polymer under the conditions used for testing. Hence, it can be concluded that tests for microwave transparent materials can be conducted in a laboratory oven.

Test conditions proposed

Based on the above results it has been possible to propose test conditions for plastic materials for use in microwave and conventional ovens (9). A distinction has been made between items which do not contact food at the point of retail sale, and which will be used by the consumer for a wide range of applications (ovenable cookware), and items which are used by industry to pre-package oven-ready foodstuffs (ovenable packaging).

If ovenable cookware is clearly labelled for microwave use with aqueous foods only, eg "not for use with foods with a high fat or sugar content", then testing for 1 hour at $100^{\circ}C$ using appropriate aqueous simulants is proposed. If the material has unrestricted

Table 3. Outline of analytical methods used to test for effects of microwave energy on migration

Polymer Type	Migration Species	Simulant	Analytical Method
PET	Oligomers (polymerisation residue)	Oil	Oligomers hydrolysed to terephthalic acid with KOH/MeOH and transmethylated to dimethylterephthalate. SEC clean up. GC-FID or GC-MS analysis. (6)
		Iso-octane	Cyclic trimer determined by HPLC.
PP	Irganox 1076 (antioxidant)	Oil & Iso-octane	^{14}C-Labelled 1076. Liquid scintillation counting.
TPX	Dilaurylthiodipropionate (antioxidant)	Oil	Hydrolysed with NaOMe/MeOH to release lauryl alcohol. SEC clean up. Derivatised with BSTFA. GC-FID or GC-MS analysis.
PVdC/PVC	Acetyltributyl citrate (plasticiser)	Oil	Stable isotope dilution method. Separated by SEC. GC-MS (SIM) analysis. (7) GC-FID
		Iso-octane	
TPES	Volatiles (manufacturing residues)	Oil	Headspace GC-MS (8)

TABLE 4. **Migration from plastics into simulants with and without microwave treatment**

Polymer	Migrant	Treatment (b)	Normalized Migration (a)	
			Approach 1	Approach 2
PET	Oligomers	NMW	100 + 14	100 + 7.1
		MW	85 + 23	114 + 21.9
PP	Irganox 1076	NMW	100 + 4.5	100 + 4.0
		MW	99 + 6.6	103 + 4.0
TPX	DLTDP	NMW	100 + 6.0	100 + 11.6
		MW	104 + 6.9	100 + 7.7
PVdC/PVC	ATBC	NMW	100 + 4.7	100 + 5.7
		MW	107 + 3.3	97 + 5.9
TPES	Benzene	NMW	100 + 3.6	—
		MW	102 + 1.2	—
	Toluene	NMW	100 + 11.8	—
		MW	86 + 7.4	—
	Ethyl-benzene	NMW	100 + 4.8	—
		MW	98 + 11.0	—

(a) Migration from non-microwaved materials normalised to 100% + S.D.
(b) MW = Microwaved, NMW = non-microwaved.

microwave oven use, it should be tested at 150°C for 30 min. For
dual-oven use, tests are to be conducted for 2 h at 175°C. The
latter two test regimes use the fatty food simulant olive oil, or
approved alternatives such as sunflower oil or the synthetic triglyceride HB307. For ovenable packaging, where the food contact
application can be specified by the manufacturer, the test temperature is selected according to the maximum temperature recorded at the
food/plastic interface when the food is cooked according to the manufacturer's instructions.

Test conditions are summarised in Table 5 and employ the recommended EC simulants of water, 3% aqueous acetic acid, 15% aqueous
ethanol and olive oil. A series of migration tests was conducted to
validate the proposed test conditions to ensure that they are more
severe than expected conditions of use but not so severe that suitable materials are excluded from use. Migration into simulants was
determined and compared to migration into real food cooked in a
microwave or conventional oven. Seven different microwaveable polymer types were used in the validation exercise (9).

For example, 0.15 mg/dm^2 of the plasticiser, acetyltributyl citrate, migrated from VdC/VC copolymer film into distilled water over
1 h at 100°C. Similar values were determined in a chicken and
sweetcorn soup when the film was used as a splash-cover during cooking in the microwave. However, when the film was used to cover a
reheated roast beef meal, migration was higher at 0.6 mg/dm^2. Under
the proposed testing scheme, VdC/VC film is classified as ovenable
cookware and has been sold in the UK for unrestricted microwave oven
use. Hence it should be tested at 150°C. However, the film will
not physically withstand such temperatures and so was tested with
olive oil for 30 min at 121°C. Under the EC test scheme (10), a
reduction factor of 5 would be applied to the migration value obtained using olive oil, when testing for contact with foods such as
biscuits and bread. The level of ATBC migration, after application
of this reduction factor, was 1.2 mg/dm^2. This is 4 times lower than
migration levels determined for biscuits, where the film was used to
line the cooking dish but twice the value for bread, where the film
was used to wrap dough.

When PET ovenable cookware was heated with olive oil for 30 min
at 150°C, 1.2 mg/dm^2 of total oligomers migrated. In comparison,
when lasagne and curry were cooked in PET, migration levels were 10-
fold lower. The appropriate reduction factor that relates to testing of these food types with fatty food simulant is three (10) and
so the final test result is 3-fold higher than the result for food.

These examples from the validation experiments serve to show that
the scheme can be used to determine migration from plastics intended
to be used at high temperatures and that the results bear an acceptable relationship to real food situations.

Research-orientated investigations to assess migration

Standardised testing schemes bring benefits of compatibility between
countries and organisations; the general suitability of materials can
be assessed through overall migration limits and safety controls can
be applied through specific migration limits. However, specific
migration limits are not a complete guarantee of safety because they

Table 5. Proposed test conditions for high temperature materials and articles

1. Ovenable cookware

Use	Test conditions	Simulants
Aqueous foods – microwave only	100°C for 60 min	aqueous
Unrestricted microwave use	150°C for 30 min	aqueous and olive oil
Unrestricted dual-oven use	175°C for 120 min	aqueous and olive oil

2. Ovenable packaging

Test according to the temperature reached during the intended use of the article.

Temp. reached in use	Test conditions	Simulants
<121°C	see EEC, 1982 (4)	As appropriate
121–150°C	150°C for 30 min	aqueous and olive oil
>150°C	175°C for 120 min	aqueous and olive oil

Note:-
(i)   If testing with olive oil can be demonstrated to be inappropriate due to the physical properties of the polymer, then there can be exclusion from the above testing, provided data can be supplied on migration into real foods.
(ii)   Testing with aqueous simulants is to be conducted at 100°C.
(iii)   For closely defined applications, the test temperature can be set at the temperature of actual use rather than at the next highest temperature bracket.

presuppose that the identity of substances that are of concern, are
all known. Even if a battery of test procedures were used to screen
for positive list compounds, there is no guarantee that they will
detect all contaminants and, potentially, these non positive-list
compounds could be of more interest from a food safety viewpoint.
Hence, in any testing scheme it is important to incorporate an element of flexibility so that adventitious contamination can be
detected.

This is illustrated by the finding of benzene at mg/kg levels
when thermoset polyester cookware articles were under examination
for emission of volatile species using headspace GC/MS techniques
(8). Investigations determined that the initiator t-butyl perbenz-
oate, used in the polymerisation process, was the source. It was
previously recognised that benzene was a major breakdown product of
this initiator, but the post-cure period of heating the finished
article was assumed sufficient to purge the residue. Testing of the
polymer solely for levels of positive list volatiles would not necessarily have detected this contaminant – indeed the presence of
benzene had escaped the manufacturer's routine analysis for other
volatile aromatics such as styrene and ethyl benzene. Flexibility
in analytical methods also permitted migration of benzene from the
cookware into simulants and real foods to be determined. Although
levels of benzene as high as 5.6 mg/kg were detected when measuring
migration from these articles into olive oil for 2 hours at 175°C in
a closed system, under more realistic conditions of cooking actual
foods, such as a meat casserole, the highest level of migration was
0.08 mg/kg. This substantiated previous observations that simulants
are generally more rigorous extractants than foods. It is also possible that there was loss through volatilisation during cooking.
Residual benzene was reduced to insignificant levels by a switch to
alternative peroxide catalysts (8).

Another example where standardised migration testing was inappropriate was where the maximum dietary intake of the plasticiser,
acetyltributyl citrate, was under examination (11). This necessitated testing of PVdC/PVC copolymer film with a wide range of foodstuffs in a variety of food contact situations. Foods examined
ranged from Swiss-roll, sandwiches and grapefruit to microwave
cooked meals such as pizza and chicken breasts. A stable isotope
dilution technique, employing clean-up of ATBC by size-exclusion
chromatography (SEC), was developed (7) to provide the necessary
sensitivity and selectivity for the determinations. Measurement of
migration into the foodstuffs allowed overall dietary exposure to
ATBC to be estimated and specific instances of exceptionally high
migration to be identified. These measurements would not have been
possible using a standardised testing scheme and simulants.

Susceptor materials provide an example where a number of potential migrants are generated during the heating process. Many of the
substances liberated are not constituents of the susceptor per se,
but are degradation products formed at the high temperature of use
(12). Compositional testing of susceptor materials as supplied at
the point of sale is therefore not adequate. Testing requires spec-
ialised techniques which are at present under study in this and other
laboratories (13, 14, 15).

Looking to the near future, the PET layer in susceptors may be replaced by other grades of PET with improved barrier properties or by polymer films such as polyetherimide. Greater temperature control may be achieved by metalizing with stainless steel or other metals in place of aluminium. Apart from susceptors, other novel materials such as polyphenylene oxide blended with polystyrene, and the wider use of co-extruded materials for retortable and aseptic packaging, could all have an impact in the microwave food contact area. Whilst a formal testing scheme can be used to assess the suitablility of materials for food contact use, such schemes must be accompanied by complementary research-oriented techniques possessing the versatility to be applied to a range of investigations. Only by the use of such a dual-approach can possible problems be identified and solved.

Literature cited

1. L Rossi, (1988) Food Addit. Contam., 5, 543.
2. L Rossi, (1988) Food Addit. Contam., 5, 21.
3. EC (1990) European Community Council Directive No. 90/128/EEC Official Journal of the European Communities L75/19.
4. EC (1982) European Community Council Directive No. 82/711/EEC Official Journal of the European Communities L297/264.
5. Anon (1987) The Materials and Articles in Contact with Food Regulations 1987. Statutory Instrument 1987 No. 1523.
6. L Castle, A Mayo, C Crews and J Gilbert, (1989) J Food Protect., 52, 337.
7. L Castle, J Gilbert, S M Jickells and J W Gramshaw. (1988) J Chromatogr., 437, 281.
8. S M Jickells, C Crews, L Castle and J Gilbert. (1990) Food Addit. Contam., 7, 29.
9. L Castle, S M Jickells, J Gilbert and N Harrison. (1990) Food Addit. Contam., 7, 779.
10. EC (1985) European Community Council Directive 85/572/EEC Official Journal of the European Communities L372/14.
11. L Castle, S M Jickells, M Sharman, J W Gramshaw and J Gilbert. (1988) J Food Protect., 51, 916.
12. J L Booker and M A Friese. (1989) Food Technol., May, 110.
13. T H Begley and H C Hollifield (1989) J Assoc. Off. Anal. Chem., 72, 468.
14. T H Begley and H C Hollifield (1990) J Agric. Food Chem., 38, 145.
15. T H Begley, J L Dennison and H C Hollifield. (1990) Food Addit. Contam., 7, 797.
16. IEC 705 (1988). Bureau Central de la Commission Electrotechnique Internationale, Geneva.

RECEIVED March 11, 1991

Chapter 3

Food and Drug Administration Studies of High-Temperature Food Packaging

Henry C. Hollifield

Indirect Additives Section, Division of Food Chemistry and Technology, Center for Food Safety and Applied Nutrition, Food and Drug Administration, 200 C Street, S.W., Washington, DC 20204

Food/package interactions that result from the use of microwave susceptors and dual ovenable trays have been extensively studied by FDA's Indirect Additives Laboratory. Temperatures in excess of 212°F are common for such packaging. Susceptors are typically constructed of plastic such as polyethylene terephthalate (PET), adhesives, and paper. Dual ovenable packaging may be constructed of crystallized PET or PET-laminated paperboard and may or may not include adhesive. At elevated temperatures, both volatile and nonvolatile chemicals from these materials can migrate into foods. An analytical procedure based on headspace gas chromatography and mass selective detection is used to determine volatiles, whereas high-performance liquid chromatography with UV detection is quite suitable for determining PET oligomers and adhesive plasticizers. A general scheme for conducting high temperature cooking simulations and migration studies is described and a summary of typical migration data is presented.

When the Food and Drug Administration (FDA) initially developed packaging regulations and guidelines, it was not envisioned that packaging materials would ever become cooking utensils or experience the temperature extremes that some package components encounter today (e.g., 400-500°F) (1). Guidelines were established for using packaging materials, such as polyethylene terephthalate (PET), based on migration data generated at temperatures no higher than about 250°F (2). Because much greater amounts of packaging components and their degradation products may migrate to foods at high temperatures, FDA has asked industry to generate data to be used in support of regulations for high temperature uses of certain packaging components (3). FDA's Indirect Additives Laboratory has conducted a number of studies on food/package interactions at elevated temperatures and their effect on food packaging, including microwave susceptors and dual ovenable trays. This paper will focus on newly developed analytical protocols for testing microwave susceptor packaging materials at elevated temperatures; however, many of these procedures are also useful in evaluating the performance of other packaging such as microwave only or dual ovenable products for which cooking temperatures are generally less severe.

Although the microwave susceptor was introduced only a few short years ago, its use has grown so rapidly that it now represents over 10% of all microwave packaging. A variety of foods are sold in microwave susceptor packaging, including popcorn, pizza, waffles, pot pies, and french fries. This market is expanding steadily into new areas of application.

The microwave susceptor package is an active cooking utensil constructed of a microwave interactive metalized plastic film that is typically laminated with adhesive to a paper or paperboard backing. This package creation can reach temperatures of 400-500°F and actually cook foods contained within it. Obviously there are questions about the performance of such a package, and data are needed to validate the integrity of the construction materials under conditions of use. First, there are questions about the formation of degradation products when the paper, adhesive, and plastic construction materials are exposed to high temperatures. Second, there are questions about the loss of package integrity at high temperatures and the attendant loss of barrier properties, resulting in increased migration of the package additives and decomposition products to food. Finally, there is concern for the possible reaction of susceptor migrants with foods to form alteration products of unknown toxicity.

To address these concerns, FDA has conducted laboratory studies on representative microwave susceptor-packaged products purchased from local supermarkets and on virgin microwave susceptor materials obtained from manufacturers. These studies have focused on several areas: determination of volatile decomposition products and additives, determination of ultraviolet light-absorbing nonvolatile extractives in food simulants and in foods, and comparison of temperatures in the package wall with those at the food/package interface.

Physical Observations

Several forms of microwave susceptors are in use today, but most can be classified as trilaminate microwave susceptor bags, like popcorn bags, or bilaminate microwave susceptor boards like those used for pizzas. The bags contain the susceptor cemented between two layers of paper, one of which is coated with a grease-resistant coating which acts as the food contact surface. The susceptor is typically a lightly metalized PET film. The bilaminated boards generally consist of a food contact surface of plastic film, most commonly PET, the underside of which has been lightly metalized. This surface is laminated to paper or paperboard by means of an adhesive layer. Other plastic films such as polyetherimides or modified polyesters have been substituted for PET. The adhesives used in this application are generally acrylate- or vinyl acetate-based. The paper component varies with the manufacturing source. Other susceptor packaging may vary in structure but typically also contains paper, plastic, and adhesive components. When subjected to microwave radiation, these susceptors get very hot, achieving temperatures required to pop popcorn, brown or crisp food surfaces, and cook pot pie crusts. Some measurements have indicated temperatures above 500°F. Photomicrographs have shown evidence of occasional melting, cracking, and breaking of the polymer food contact surface. Because of the fragile nature of the materials used to construct the susceptor, it is easily abused at high temperatures by overheating for only a minute or two.

Heating of microwave susceptor materials results in the release of large numbers of volatile and nonvolatile substances having a wide range of chemical functionalities. Therefore, several techniques have been required for measurement of the migrating substances. Headspace gas chromatography (GC) with flame ionization and mass spectrometric (MS) detection has been used to determine the more volatile components. High-performance liquid chromatography (HPLC) with UV detection has been extensively used to determine UV-absorbing nonvolatiles.

Volatiles

A general headspace gas chromatographic analytical procedure has been developed for determining the volatile substances emitted from high temperature food packaging (4). It has been used to screen microwave susceptors and fatty food simulants. Foods are not useful test vehicles in this case because they generally contain too many natural volatile components that interfere with the analysis.

Figure 1 represents a chromatogram of the volatile substances emitted from a portion of virgin susceptor stock that had not contacted a food product. The chromatogram was obtained by placing a 1 sq. in. strip of microwave susceptor in a 22 mL headspace vial and sealing it with a Teflon-lined silicone rubber septum and an aluminum crimp cap. The vial was microwaved with 250 mL of water in the oven to act as a microwave load. Microwave exposure times were chosen so that after heating, the test susceptor strips would be similar in appearance to a susceptor that had been actually used to cook the food. The microwave oven was a typical 700 W consumer model available at most local department stores. After microwave heating of the susceptor, the resulting volatiles retained in the sealed vial were analyzed by headspace GC.

Two instruments were used in this study. The first, which was used primarily for quantitation, was a Perkin-Elmer Sigma 2000 gas chromatograph equipped with an HS-100 automated headspace sampler and a flame ionization detector. A 30 m, 0.25 mm id, 5% phenylmethyl silicone capillary column gave adequate resolution. The instrument was also equipped for subambient oven operation and had a purge/trap interface for measurement of very volatile compounds. Qualitative analysis was performed by a Hewlett-Packard 5890A gas chromatograph equipped with an HP 5970B mass selective detector and a 5% phenylmethyl silicone capillary column.

The headspace procedure described above allows comparison of all susceptor products under the same conditions and determination of the relative amounts of the volatile chemicals produced. This procedure has been used to survey microwave susceptor products currently available in local supermarkets. In a recent survey, 11 unique susceptor products were analyzed. They varied in composition and in the number and amounts of substances evolved. Some of the volatile substances identified may have come from the food products, which were in contact with some of the susceptors. Seven susceptor products contained acrylic-based adhesives; the remainder contained vinyl acetate-based adhesives.

Generally, the more paper in the susceptor construction, the greater the amounts and number of volatile substances released on heating. The number of substances released from a given susceptor ranged from 34 to 105. Compounds with boiling points from about 34 to 234°C were represented. Some of the identified chemicals that were emitted at levels greater than 0.5 μg/sq. in. of susceptor surface are listed below:

acetone   furan
methyl vinyl ketone   isobutanol
1,1,1-trichloroethane   acetic acid
crotonaldehyde   butanol
pentanal   methyl methacrylate
toluene   3-methyleneheptane
hexanal   2-propoxyethanol
furfural   styrene
heptanal   2-ethyl-1-hexanal
benzaldehyde   octanal
2-ethyl-1-hexanol   1-phenylpropanedione
nonanal   octyl acetate
butoxyethoxyethanol   5-hydroxymethylfurfural

This level corresponds to a concentration of 50 ppb when 10 g of food is in contact with a surface area of 1 sq. in. Some of the identified chemicals found at levels lower than 0.5 μg/sq. in. are listed below:

isopropanol   butanal
vinyl acetate   methyl ethyl ketone
formic acid   benzene
methyl furan   butanoic acid
furan methanol   butyl ether
butyl acrylate   2-heptanone
3-heptanone   2-butoxyethanol
o-dichlorobenzene   benzoic acid
decanal

These volatile substances encompass a wide variety of chemical functionalities, including numerous aldehydes, ketones, alcohols, and carboxylic acids. Paper and adhesive components and their degradation products are the major sources of the volatile chemicals. Although most are emitted in small amounts, a few are released in amounts that would correspond to parts-per-million levels in foods if the total amounts migrated. The cumulative amount of each chemical available to migrate to food varies, depending on the susceptor construction, cooking time, and maximum temperature achieved. Migration studies using corn oil have demonstrated that volatile substances such as furfural, 2-butoxy-1-ethanol, and benzaldehyde are retained in hot vegetable oil after cooking.

Any attempt to quantify these volatile substances is handicapped by several factors. For example, the early eluting chemicals can be separated only with great difficulty. Use of a subambient GC oven is helpful in performing this task. Even so, the abundance of free carboxylic acids, such as acetic acid, in some susceptor products interferes with the resolution of some substances. Because of the strong adsorption of polar chemicals by the

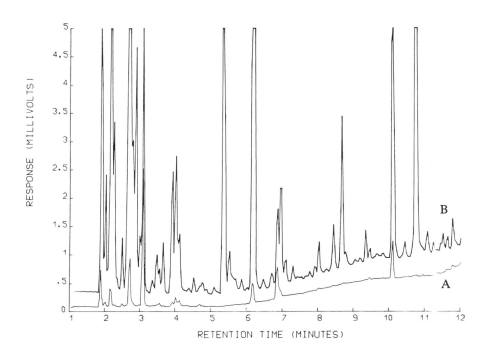

Figure 1.   Headspace gas chromatograms of microwave susceptor volatiles (A) before and (B) after exposure to microwave heating.

susceptor cellulose, the procedure may not be very sensitive for determining small amounts of many substances. Also, water, which may be present in the susceptor at levels up to 12%, can have two opposing influences on the determination. First, it can lessen the adsorption of polar substances on the susceptor paper and, second, if present in the condensed phase in the headspace vial, may solubilize the polar substances and reduce the sensitivity of the headspace procedure for these chemicals.

Because of these competing factors, it is difficult to quantify many of the polar volatiles generated from the degradation of the adhesive and paper materials. Attempts to conduct recovery studies using 4-heptanone as an internal standard have resulted in large errors in measurement of some chemicals such as polar oxygenated compounds like 2-furanmethanol. So far, the only quantitative technique found to avoid this problem has been the use of standard additions. This solution assumes that the GC peak has been identified and that a standard has been obtained. Obviously, a problem remains when standards are not available or the peak cannot be identified.

Nonvolatiles

Several investigators have made important contributions to the determination of PET oligomers and residues that migrate from food packaging to foods and simulants (5-8). However, most of these methods measure terephthalic acid esters and are not specific for PET oligomers or other package migrants. FDA developed the first more general HPLC method for the determination of specific PET oligomers and UV-absorbing compounds such as adhesive components from microwave susceptors. Subsequently, this methodology has been successfully applied to both the package materials and the foods prepared in them (9-12).

Potential migrants of higher molecular weight, such as plasticizers and oligomers, have been arbitrarily classified as "nonvolatiles" and are generally determined by HPLC. However, some are certainly volatile enough to be determined by GC, and are referred to as "semivolatiles." In fact, a GC/MS total ion chromatogram of an acetonitrile extract of a pot pie-type susceptor construction reveals about 35 individual peaks of components ranging from C16 hydrocarbons to the PET cyclic trimer (Figure 2). Most of these substances are not extracted before the susceptor is heated in the microwave, suggesting that many are probable degradation products. So far, HPLC methods have been developed for monitoring the migration into foods and food simulants of only the UV-absorbing substances. Attempts to monitor migration of the non-UV-absorbing migrants have not met with much success because of the poor sensitivity and

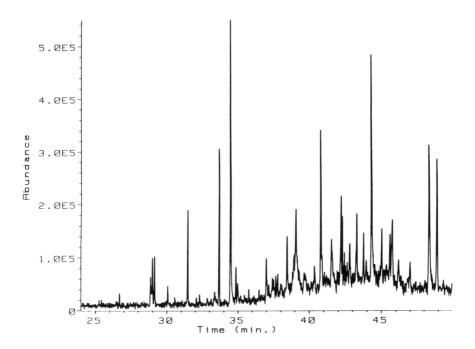

Figure 2. GC/MS total ion chromatogram of a concentrated acetonitrile extract of a previously heated microwave susceptor showing semivolatile components eluting before the PET cyclic trimer.

specificity of general HPLC detectors. Some researchers
have tried to determine total nonvolatiles and the amounts
of non-UV-absorbers by extraction and gravimetric
procedures, but these approaches have yet to be validated.

The FDA procedure for monitoring migration of UV-
absorbing nonvolatile residues from microwave susceptors
uses Miglyol, a food-simulating oil which can be exposed to
high temperatures with a minimum of degradation. Refined
naturally occurring corn oil was the original choice for
these studies because it was readily available and did not
interfere with PET oligomer determination. Initially, it
was mistakenly thought that the oligomers from the PET food
contact layer were the only major migrants of interest. It
was later learned that PET barrier properties diminished
when the film was heated above its glass transition
temperature, allowing additives and decomposition products
from the adhesive and paper susceptor components to migrate
readily through the PET to foods and food simulants. Corn
oil decomposition products interfered with the determination
of the adhesive and paper-based migrants. Other oils were
examined and several possible alternative food simulants
were identified. These included HB-307 (a synthetic
triglyceride that is widely used in Europe), paraffin oil,
and Miglyol (a distilled fraction of coconut oil). All
developed relatively few chromatographic interferences under
the test conditions used (11).

The migration experiments were conducted on unused
susceptor stock or on actual susceptor products purchased
in local supermarkets. Oil migration experiments were
performed using a special Teflon single-sided extraction
cell designed at Waldorf Corp. (St. Paul, MN). All
susceptor test materials were cut from flat areas of the
packages to fit the extraction cell. For further details,
see Ref. 11. In our cell the exposed susceptor area was
7.07 sq. in. The ratio of oil weight to surface area was
7.1 g/sq. in. The temperature of the oil during microwave
cooking was monitored with a Luxtron Model 755 fluoroptic
temperature-sensing instrument. Food oil simulants were
typically cooked for 3 min. Foods such as french fries were
cooked according to package directions, usually for the
maximum recommended time. Controls were prepared by cooking
the oil or french fries in a Pyrex petri dish heated by a
microwave susceptor board placed beneath the dish to prevent
direct contact with the susceptor. The cell was not used
for the control experiments. After microwaving, 3 g of oil
or all of the cooked french fries were extracted and
analyzed for migrants.

For extraction, an aliquot of the oil was transferred
to a separatory funnel, diluted with 50 mL of hexane, and
extracted with two 25 mL portions of acetonitrile (9, 11).

Foods were extracted in an explosion-proof Waring blender with hexane, and the extracts were filtered and extracted with acetonitrile in a similar manner (10). The acetonitrile layer was concentrated in a Kuderna-Danish evaporative concentrator and analyzed by solvent-programmed reversed-phase HPLC, using a Microsorb-C, 5 μm particle size, 250 X 4.6 mm column, an acetonitrile/water mobile phase, and a Waters Model 480 Lambda Max variable wavelength detector.

A number of susceptor products have been analyzed by this technique. Table I shows clearly that several chemicals from the susceptor migrate readily to both oil and food (french fries). The levels found in the oil are generally higher than those found in the food. The concentrations of these nonvolatile susceptor migrants appearing in foods are also significantly higher than those of the volatile chemicals. The UV-absorbing nonvolatile chemicals found in the highest concentrations are typically PET oligomers, plasticizers, and adhesive components. All of these substances also have been found in foods prepared by using susceptors. The principal residues found in several of the foods include diethylene and dipropylene glycol dibenzoates. These compounds occur in food at levels comparable to those of the PET oligomers (about 15 ppm each). The diglycidylether of bisphenol A, an epoxy compound, has also been found in some packages. The highest levels of oligomers and plasticizers found in a single food such as french fries has been about 45 ppm (PET oligomers plus the two plasticizers). However, this value does not include estimates for either volatile or non-UV-absorbing nonvolatile migrants. We have found PET oligomers and various plasticizers in foods such as pizzas, waffles, sandwiches, pot pies, and french fries at parts-per-million levels.

Table I. Corresponding Migration Levels of Nonvolatile Components in Food and Food-Simulating Vegetable Oil

Migrant	Food (ppm)	Oil (ppm)
PET oligomers	7.6	131.0
diethyleneglycol dibenzoate	11.0	131.0
dipropyleneglycol dibenzoate	7.8	90.4
diglycidylether of bisphenol A	1.5	--

In addition to the typical microwave construction described earlier, studies have been conducted on four major brands of microwave popcorn bags. Two plasticizers, diethylene glycol dibenzoate and dipropylene glycol dibenzoate, along with PET oligomers were found to migrate

into Miglyol fatty food simulant from at least one test bag.
An unidentified alcohol migrated from two other bags. It
is estimated that none of the popcorn bag migrants reached
parts-per-million levels in food.

Possible Food Alteration Products

The nonvolatile migrants from microwave susceptors are
thought to be made up of all types of chemical moieties.
Several alcohols are known to be either used as raw
materials in susceptor constructions or formed as
decomposition products of susceptor components. A procedure
for alcohols was developed to obtain more information on
these chemicals. This procedure incorporated a reaction of
dinitrobenzoyl chloride with alcohol groups in acetonitrile
extracts of fatty food simulants heated in contact with
microwave susceptors. Preliminary tests showed this approach
to be feasible. In practice, however, recoveries from
fortified Miglyol test extracts were quite low. However,
when paraffin oil fatty food simulant was substituted in
this experiment for the vegetable oil, very good recoveries
were obtained. It is possible that, at the high
temperatures achieved during microwave heating, fatty acid
ester-based foods such as vegetable oils react with alcohol
migrants by transesterification, resulting in the observed
low recoveries.

Given the large numbers of chemicals that migrate from
the microwave susceptor and the very high temperatures at
the food/package interface, the potential exists for many
food alteration products to form.

Temperature Measurements

One of the more controversial tasks associated with the
study of high temperature packaging is measuring the maximum
temperature attained by package materials. These data are
important because excessive temperatures lead to material
failure and the formation of degradation products. Also,
at elevated temperatures, chemical additives and other
residues in the package will migrate more rapidly to the
surface where they can transfer to the food. Because
internal package temperatures are not easily determined and
may vary widely from side to side or top to bottom,
scientists have tended to use food/package interface
temperature measurements to describe the heat histories of
packaging materials (12). The point is that these interface
temperatures are often lower than those attained by the
package and not truly indicative of the degradation of the
package material or of the migration of package residues to
the food product. These temperatures are particularly
variable for microwave susceptors during microwave heating.

These observations are supported by infrared photography of susceptor surfaces during microwave heating and also by differential scanning calorimetry (DSC) experiments on previously heat-treated PET films. Infrared photography of the surface of a susceptor in a microwave oven shows a constantly changing heat pattern when the susceptor is subjected to microwave fields. The temperature pattern ranges from moderate to extremely hot. The key phrase is "constantly changing heat pattern." Such patterns are not accurately portrayed by averaged temperatures obtained by point source measurements. Moreover, when food is cooked and the temperature is monitored by infrared camera, generally higher temperatures are obtained in some areas of the container than are indicated by fluoroptic probe measurements (1). Obviously, any temperature probe at the food package interface will produce an average temperature between that of the hot cooking surface and the cooler food mass that is being heated. A probe located between a very hot microwave surface and a frozen piece of breaded fish will indicate a temperature intermediate between the two. Assuming that the microwave energy source is reasonably constant, the susceptor temperature will always be hotter than the fluoroptic probe temperature. A food mass largely made up of water will appear to be at a steady state as it loses water, but the interface temperature can be significantly lower than the maximum internal package temperature.

More recently, a variation of a DSC technique used by Moore et al. (13) to study the melting behavior of amorphous PET provided data on the maximum temperature attained by the PET film. By using DSC, endotherms were obtained on both heated and unheated test samples. Unheated amorphous film has a very characteristic DSC trace. On the other hand, when it is placed between the food and package and heated in a microwave or conventional oven, its post-test DSC curve shows an endotherm indicative of increased crystallization that has occurred as a result of the heat treatment. The appearance of the film after heating typically shows a patchwork of clear and opaque areas, with the opaqueness representing increased PET crystallinity. Some areas become hotter than neighboring areas, possibly because of variations in the microwave field and in food composition and contact surface. DSC curves are selectively run on those portions of the film that display the greatest degree of crystallinity and, therefore, experience the highest temperature. For example, laboratory samples were tested in a hot air oven set at 200°C. DSC curves were taken before and after heating. The indicated maximum temperature was 200 ± 3°C.

Table II shows temperatures obtained by both fluoroptic probes and DSC curves for PET films heated in cooking

experiments. Results are given for controls heated in the
hot air oven and for susceptor cooking experiments on pizza
and pot pie. In almost every case, the maximum temperature
attained by the package (as determined by DSC) exceeded that
recorded by the fluoroptic probe. This difference suggests
that the fluoroptic probe temperatures are not
representative of the maximum internal package temperatures.

**Table II. Food-Package Interface Temperature Estimates
Obtained by Differential Scanning Calorimetry (DSC) and
Fluoroptic Probes**

Product	DSC Av. Temp. (°C)	N[a]	Fluoroptic Temp. (°C)[b]
pizza	207	4	130, 185, 140
pizza	184	1	137, 134, 143
pizza	228	1	203, 212, 207
pizza	220	6	112, 133, 156
meat pot pie	196	2	104, 176
pizza	216	3	204, 221, 219
pizza	222	3	197, 201, 210
pizza	220	5	154, 173, 174
film (150°C)[c]	154	4	
film (175°C)[c]	180	3	
film (201°C)[c]	206	2	

[a]DSC determinations were conducted N times on the same
specimen.
[b]Measured at three different locations at the food-
susceptor interface.
[c]PET films heated in a temperature-controlled hot air oven
were used for controls.

These tests are significant because they raise
questions about the appropriateness of various temperature
measurements. Which determinations are most representative
of food packaging materials during high temperature cooking
applications? Food/package interface temperature measurements that are not indicative of the internal package
temperature would not appear to be reliable or reproducible
indicators of package degradation or additive migration.
For this reason, it is felt that calculated migration data
or simulated migration tests based on the use of the
interface temperature measurements, such as time-temperature
product profiles, are not reliable substitutes for actually
measuring the migration levels.

Discussion

Methods in use at FDA for determining volatile and UV-absorbing nonvolatile substances have been described. Data have been presented to show significant migration of numerous chemical substances to foods cooked in microwave susceptors. The methods are also applicable to other high temperature food packaging, such as the dual ovenable containers used in both microwave and conventional ovens.

These observations have dealt with food/package interactions of the microwave susceptor in its various forms. The products reach very high temperatures causing paper and adhesive components to degrade and emit numerous volatile chemicals. Also, at cooking temperatures the plastic softens, may melt, and often tears apart. The number of volatile residues produced by heating 40 test materials has been reported to exceed 200 (14). It is estimated that the total number of substances emitted, when all of the nonvolatile chemicals are also accounted for, may be more than 300. Many of these substances have been identified as adhesive or paper additives or their degradation products. The concentrations of substances produced appear to be related to the paper and adhesive used in the susceptor and to the temperature achieved. Migration tests have demonstrated that the typical microwave susceptor in use today readily transfers parts-per-million amounts of plastic, paper, and adhesive additives and degradation products to the prepared foods. Fluoroptic temperature measurements and DSC data show that food/package interface temperatures vary significantly from those of the inner package material. Use of lower interface temperatures for simulated migration experiments underestimate migration levels and the extent of degradation. It seems reasonable, therefore, to conclude that actual migration studies of microwave susceptors are the only reliable means of determining the transport of additives or degradation products to foods.

Obviously, we have only begun to develop methods for monitoring migration of additives and degradation products from high temperature packaging materials. There is still much to do in this area to gather adequate data to understand how microwave susceptors and their various component materials perform in these extreme environments. Finally, there is the task of writing suitable testing protocols to ensure the integrity and safety of high temperature packaging materials used in the preparation of foods.

Literature Cited

1. Lentz, R. R.; Crossett, T. M. *Microwave World* 1988, 9(5), 11-16.

2. <u>Code of Federal Regulations</u> Title 21, Part 174-177, U.S. Government Printing Office: Washington, DC, 1989 (PET and adhesives).
3. ANPR Packaging Materials for Use Under High Temperatue Conditions in Microwave Ovens. <u>Fed. Regist.</u> Sept 8, 1989, <u>54</u>, 37340-37343.
4. McNeal, T. P. <u>FDA Protocol for Semi-Quantitative Determination of Microwave Susceptor-Derived Volatile Chemicals, 1989.</u> Available from the Food and Drug Administration, Division of Food and Color Additives, HFF-330, 200 C St., S.W., Wash., DC 20204.
5. Castle, L.; Mayo A.; Crews C.; Gilbert, J. <u>J. Food Prot</u>. 1989, <u>52</u>, 337-342.
6. Tice, P. A. <u>Food Addit. Contam.</u> 1988, <u>5</u>, Suppl. 1, 373-380.
7. Ashby, R. <u>Food Addit. Contam.</u> 1988, <u>5</u>, Suppl. 1, 485-492.
8. Hudgins, W. R.; Theurer, K.; Mariani, T. <u>J. Appl. Polym. Sci. Appl. Polym. Symp</u>. 1978, <u>34</u>, 145-155.
9. Begley, T. H.; Hollifield, H. C. <u>J. Agric. Food Chem</u>. 1989, <u>38</u>, 145-148.
10. Begley, T. H.; Dennison, J. L.; Hollifield, H. C. <u>Food Addit. Contam</u>. 1990, <u>7</u>, 797-803.
11. Begley, T. H.; Hollifield, H. C. <u>J. Food Prot</u>. 1990, 53, 1062-1066.
12. Kashtock, M. E.; Wurtz, C. B.; Hamlin, R. N. <u>J. Packaging Technol</u>., 1990 (manuscript submitted).
13. Moore, J. T.; Reeves, B. J.; Rothwell, L. A. Eastman Kodak Co. Monograph, 1989.
14. Comments of the NFPA-SPI Susceptor Microwave Packaging Committee (Docket No. 89N-0138, available from the Hearing Clerk) Re: Advance Notice of Proposed Rulemaking Regarding Packaging Materials for Use Under High Temperature Conditions in Microwave Ovens. <u>Fed. Regist</u>. Sept. 8, 1989, <u>54</u>, 37340.

RECEIVED April 26, 1991

Chapter 4

Interactions of Food, Drug, and Cosmetic Dyes with Nylon and Other Polyamides

L. L. Oehrl[1,3], C. P. Malone[2], and R. W. Keown[3]

[1]Southern Testing and Research Laboratories, Inc., 3709 Airport Drive, Wilson, NC 27893
[2]Department of Textiles, Design, and Consumer Economics and
[3]Department of Food Science, University of Delaware, Newark, DE 19716

The Food, Drug and Cosmetic (FD&C) dyes are all low molecular weight acid dyes containing sulfonic acid salt auxochromes which result in interactions with polymer systems containing complementary functionalities capable of reacting to form chemical bonds. These include bonding between the amine in nylon and proteins and the group on the dye molecule. These interactions present problems in food systems using these dyes due to migration of the color between foodstuffs, packages and other non-food materials. The factors affecting the sorption of dyes from solutions encompassing concentration levels in processed foods by nylon and other polyamides are reported. This paper reports investigations into the mechanisms and kinetics of the interactions between the dyes and polymers. The effects of pH, time, temperature and dye concentration were examined. Adjuncts common to foods were investigated as to their ability to change the dye to substrate interactions.

The appearance of a food product is considered one of the most crucial components of its acceptance by the consumer. The color of foodstuffs defines their appearance. The importance of preconcieved ideas about the colors of these materials is most notable in processed foods. They may be colored with natural materials but more often rely on synthetic colors to create the desired effects.
 The uses of these colors are primarily in highly processed foods such as soft drinks, baked goods,

0097–6156/91/0473–0037$06.00/0
© 1991 American Chemical Society

candies and dry mixes for cakes, gelatins, puddings and drinks (1). The majority of other types of foods which require a limited amount of processing are not major users of these colors. It is these highly processed foods which also are most frequently packaged with various types of plastics. Due to the frequency of contacts between synthetic food colors and plastic materials, the problems of their interactions need to be addressed.

Due to good gas barrier and grease resistance properties, nylon and nylon containing films have been used extensively for the vacuum packaging of meats and dairy products (2). In addition they have excellent toughness. They maintain these good properties in the presence of a number of environmental stresses such as temperature. Some observations have been made that when foods containing dye materials are exposed to polyamides, of the nylon 6 variety, scalping of the dyes can occur.

Consumer acceptance plays a vital role in the success of a food product. Bleeding of colors between food components creates an unappetizing appearance. Migration of colorants into a package gives a dingy and unsanitary looking product. Staining of household materials by processed foods and beverages containing the synthetic food dyes generates complaints by their users and leads to decreased product sales. In order to limit such problems, knowledge of the interactions between food colorants and both food and nonfood materials is essential.

The present scientific literature contains almost no information regarding these types of studies. Much of the usage of food dyes is empirical and therefore organized discussion on these interactions is lacking.

There are eight compounds that carry the F D and C designation (Table I). Our initial response upon first examining the dye compounds was to group them based on their molecular structures. There are two triphenylmethanes (Blue 1 and Green 3), one indigoid (Blue 2), one xanthene (Red 3), three monoazos (Red 4, Red 40, Yellow 6) and one pyrozolone (Yellow 5). Since Red number 4 is no longer permitted in food materials, and Red number 3 was found to be chemically very different, they were not part of this study.

As the research progressed, it was found that the common moiety between all these dyes, the sulfonic acid auxochrome attached in each case to a phenyl ring, was the most crucial element in dye to polymer interactions so that structures are represented generically as seen here.

$$DYE - \langle\!\langle\ \rangle\!\rangle - SO_3^-$$

Table I. FD&C COLORANTS

Brilliant Blue (FD&C Blue No 1.)
Fast Green FDF (FD&C Green No. 3)

Indigo Disulfoacid (FD&C Blue No. 2)

Erythrosine (FD&C Red No. 3)

Ponceau SX (FD&C Red No. 4)
Allura Red (FD&C Red No. 40)
Sunset Yellow (FD&C Yellow No. 6)

Tartrazine (FD&C Yellow No. 5)

The substrates were chosen based on two criteria
(Table II). The frequency of their contact with F D
and C dyes and the presence of protonatable amine
groups on the molecule. The packaging polymer was
Nylon 6.6. For common household surfaces, melamine
formaldehyde resin commonly known by its trade name
Formica and the proteins wool and silk were used. As
the food model, texturized soy protein was used. The
nylon, wool and silk substrates were exposed to dye
solutions in the form of fiber skeins. Although there
are comparison problems between fibers and films due to
differences in surface to volume ratios, the skeins
were easier to handle and adequately served to
demonstrate the effects of dyes on the substrates.

Table II. Substrates

Nylon 6.6 $\quad$ $H-[-HN(CH_2)_6NHCO(CH_2)_4CO-]_n-OH$

Proteins $\quad$ NH_3^+-Amino Acids-COO

Melamine Formaldehyde
Resin

$$CH_2OH$$
$$|$$
$$NH$$
$$|$$

METHODOLOGY.

Colorants. Dyes bearing the F D and C designation were purchased from commercial sources in the powder form as their sodium salts. F D and C Green No. 3 was purchased from Tricon Colors (Elmwood, NJ). The remaining colors were purchased from Warner Jenkinson (St. Louis, MO). Stock solutions were prepared in distilled water at concentrations of 2000 parts per million (ppm) dye based on the percentage dye (or purity of the dye listed in the product literature). New solutions were prepared on a monthly basis and checked before useage for degradation by visible spectroscopy.

Substrates. Using standard five gram samples of the fibers called skeins, the substrates were exposed to solutions of the dyes using an Ahiba WBRG 60 Texomat (Ahiba, Inc., Charlotte, NC). This apparatus is commonly used in textile science and allows for the dyeing of fibers under controlled conditions of temperature and agitation. It was also used for the Formica substrate by cutting the material into strips and attaching the strips to the skein holders. The soy protein used was in the form of lumps that approximate a one inch dice normally seen in meats. After rehydrating the dried protein in water, normal laboratory beakers and magnetic stirrers were used to expose this material to the dye.

pH Variation. For these studies, 20 ppm solutions of the dye were prepared. This concentration is similar to the concentration of colorants found when commercially prepared instant beverage mixes are diluted for use according to package instructions. The pH of the solutions was adjusted in 1 unit increments over a range of 2 to 8. This is the pH range of almost all foodstuffs. The pH adjustments were made using 1.0 $\underline{M}$ citric acid or sodium hydroxide solutions. Substrates were exposed to the solutions for a period of 1 hour under controlled temperatures.

Concentration Variation. Solutions of the dyes were prepared by serially diluting an aqueous stock solution of 2000 ppm dye. The range of concentrations encompassed 0.2 ppm to 2000 ppm with at least two points in each 10-fold dilution. A total of 5 concentrations were used over the range commonly seen in powdered beverage products diluted to their consumable strengths (10 to 50 ppm). The solutions were adjusted to pH 3.0 with 1.0 $\underline{M}$ citric acid. The substrates were exposed for one hour to these solution at 25°C.

Temperature Effects. Solutions of dye were prepared as in the pH variation experiments. Substrates were exposed to these solutions for 1 hour at 2^c, 25^c, 50^c and 100^c C.

Rate Determination. Solutions of the dyes were prepared at concentrations of 20 ppm and adjusted to pH 3.0 with 1.0 $\underline{M}$ citric acid. A series of nylon and wool substrates were exposed to the solutions at 100^c, 50^c and 25^c C for time periods ranging from 1 second to 1 hour. The action of the dye adsorption was stopped by plunging the substrates into ice baths.

Mechanism Determination. Five solutions of 20 ppm F D and C Blue No. 1 were prepared from the stock solution. The pH of each solution was adjusted to 3.0 with one of these acids, trichloroacetic acid (pKa 0.52), oxalic acid (pKa 1.27), sulfuric acid (pKa 3.00), citric acid (pKa 3.13) and acetic acid (pKa 4.76). Nylon substrates were exposed to these solutions for 1 hour at 25^c C.

Additive Effects. Solutions of 20 ppm dye were prepared from stock solutions and the pH of each solution was adjusted to pH 3.0 with 1.0 $\underline{M}$ citric acid. Sodium choride or sodium sulfate crystals were added to give a series of solutions over a concentration range of 1 μM to 1.0 $\underline{M}$ salt. In another series of experiments, crystals of fructose, glucose or sucrose were added to give a series of solutions with a range of concentration of 0.1 to 1.0 $\underline{M}$ sugar.

Quantitation. After the dye exposure, the substrates were rinsed with water, air dried and the color change was measured using a handheld colorimeter. The resulting Hunter L*a*b* values were quantitated as described in the results section.

Using a visible spectrophotometer, a measurement was made of each dye solution before introduction of the substrate at the characteristic absorbance maximum for the individual color. After the trial was complete, the maximum was again measured. The change in absorbance of the colorant solution was quantitated as discussed below.

RESULTS AND DISCUSSION

pH Variation. The variations of the pH of the dye solutions indicated (Figure 1), that the chemical composition of a food in regards to pH and protonable amine content had a significant effect on the coloration of the food by the F D and C dyes and on migration of the dye from the food to other materials. With the system soy protein in 20 ppm solutions of Blue

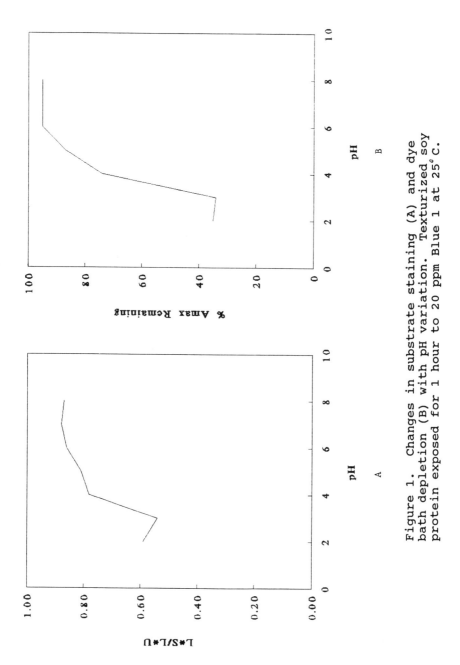

Figure 1. Changes in substrate staining (A) and dye bath depletion (B) with pH variation. Texturized soy protein exposed for 1 hour to 20 ppm Blue 1 at 25° C.

1, at one hour exposure times, the greatest dye uptake
by the substrate occurred between pH three and four.
This change was examined by following two parameters.
The first was to monitor the color change on the
substrate (Figure 1A). Using the Hunter L* a* b*
system, the variation of L* was followed as this
measure. The ordinate scale is the numerical
comparison of L* values for the substrate before and
after exposure to a dye containing solution.
The second method was to monitor the absorbance
maximum of the dye solution itself using visible
spectroscopy. Each dye has a characteristic absorbance
maximum value which is directly related to the
concentration of the dye in solution. Comparison of
the dye remaining in solution after exposure of a
substrate to the original concentration of the dye is
seen in Figure 1B. From this, it is seen that at pH
3.0, 65 percent of the dye was removed from the
solution but at pH 8, only 5 percent was absorbed.
From either of these two measurements, it was observed
that as the pH of the solution increased, the staining
of the substrate and the dye uptake decreased. This
pattern was found to hold true regardless of the F D
and C dye used and the amine containing substrate
monitored.
        In a system where the nylon package comes into
contact with dye containing food, this phenomena can
result in package staining. The system of exposing
nylon to dye containing solution modelled that of
colored beverages packaged in nylon bottles. The same
pattern of staining was seen both in dye uptake by the
nylon and in color intensity of the fibers after
exposure (Figure 2).
Concentration Variation. By varying the concentration
of the dye solution, it was found that the dye uptake
was dependent on the number of protonatable amine ends
on the substrate. The variation was in effect, a
titration of the end groups (Figure 3). Dye uptake
continued, dropping rapidly as saturation was
approached. These titrations were quantitated in terms
of the literature values of the number of amine ends on
the substrates. The pattern of the data in Table III
shows that as the number of amine ends increases, so
does the amount of dye depleted from the solution.
        From the same study, dye uptake was seen to be
dependent on the number of sulfonic acid groups present
on the dye (Table IV). Nylon bound less Blue 1 than
any of the colors containing two sulfonics.
Examination of microscopic cross sections of stained
nylon and wool substrates revealed that even at the
earliest time points, the colorants had migrated
completely through the fibers. Diffusion through the
wool substrate was faster than through nylon due to
wool's greater moisture content. It will reach
equilibrium swelling faster than nylon in water

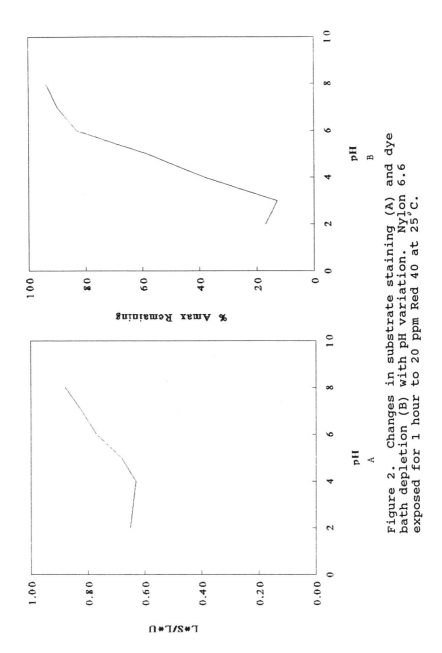

Figure 2. Changes in substrate staining (A) and dye bath depletion (B) with pH variation. Nylon 6.6 exposed for 1 hour to 20 ppm Red 40 at 25°C.

TABLE III. Blue 1 Uptake By 5 Grams Fiber, 1 Hour, 50^0 C

Substrate	Mol/l Dye Depleted	Milli Equivalents Amine Ends	
Nylon 6.6	6.5 x 10-5	0.037	(3)
Silk	9.0 x 10-4	0.15	(4)
Wool	1.3 x 10-3	0.80	(5)

solutions due to the presence of hydrophlic amino acid side chains. Equilibrium absorption of colorant is achieved by both substrates after about 15 minutes. After 1 hour, the differences between the dye binding were attributed to more binding sites being occupied by the molecule due to the larger number of sites able to bind.

TABLE IV. Dye Uptake By 5 Grams Nylon Fiber, 20ppm Dye, 1 Hour, 50^0 C

Color	Sulfonic Groups	Mol/l Dye Depleted
Blue 1	3	6.5 x 10-5
Blue 2	2	1.8 x 10-3
Red 40	2	4.5 x 10-4
Yellow 5	2	1.9 x 10-4

Temperature Effects. In the chemistry of textile dyeing, it has long been known that the most efficient conditions for coloration of fibers are at boiling temperatures (5). This research has confirmed that the intensity of staining by food dyes increased with increasing temperatures but staining was still visible

at low temperatures if the pH conditions were low
enough (Figure 4). Even at ice temperatures, nylon was
seen to pick up 26 percent of the dye in solution,
while at room temperature, 87 percent was removed at
low pH values. While the small decrease in the dye
content of the bath was not found to be statiscally
significant in numerical terms, the overall trend of
the data is followed at ice temperatures. The
corresponding change of color of the nylon substrate is
visible to the eye, indicating that a low temperature
does not offer total protection from the exchange of
dye from the solution to the nylon substrate. As the
number of available amine ends on the substrate
increases, the amount of dye depleted from the solution
increases as outlined in Table III. With wool, nearly
90 percent of the dye was removed from a bath of Red 40
at $2°C$. At these low temperatures, plastic packaging
may be relatively safe from dye migration but exchange
of color between other proteins in a food product or
between food product and household surfaces will be
actively occurring.

Rate Determination. The rate of staining was also
found to increase with increasing temperature.
Diffusion of the dyes into all the polymers using the
system was very fast with total saturation occurring in
most cases between 5 and 10 minutes. Using the wool
substrate as an example, an Arrenhius plot of the
temperature dependence of the rate constants is shown
in Figure 5 for the two blue colors. From this plot,
the activation energies by dye depletion were
calculated to be 42 kilo joules per mole for Blue 1 and
71 kilo joules per mole for Blue 2. The relative
closeness of these values indicated the mechanism of
color adsorption was the same despite the differences
in chemical structure between the two colors.

Mechanism Determination. In conjunction with the pH
studies, the relationship of the pKa of the acid used
to make the pH adjustment to the color uptake by the
nylon substrate was studied in order to define the
mechanism. The stronger the acid, the less required to
make the appropriate pH change to three. This resulted
in less protons available to ionize the nylon amine end
groups. As the number of protons in the bath
increased, more ionization occurred and more dye anions
were removed from the bath to maintain
electroneutrality on the fiber. This resulted in the
greatest dye bath depletion when weak acids were used
(Figure 6). It has been proposed the weak acids are
absorbed by nylon to a greater extent than strong ones
at the same pH in an undissociated form by hydrogen
bonding to the amide groups. This would enable sites
other than terminal ones to bind with dye molecules
(5).

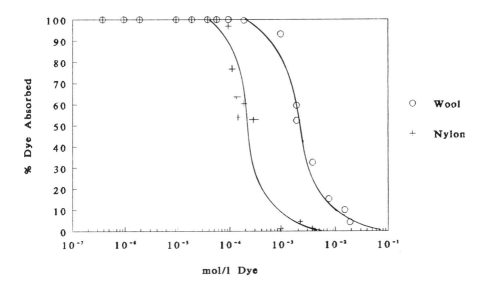

Figure 3. Dye uptake by substrates as a function of dye bath concentration. Substrates exposed to solutions of Blue 1 for 1 hour at 50° C.

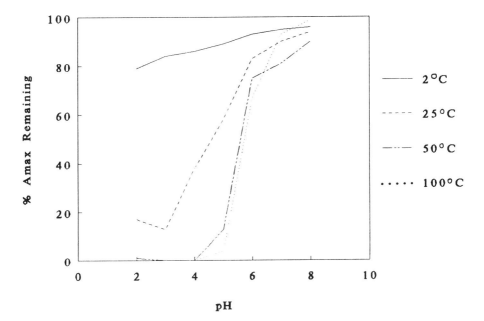

Figure 4. Changes in dye bath depletion with pH at various temperatures. Nylon 6.6 exposed to 20 ppm Red for 1 hour.

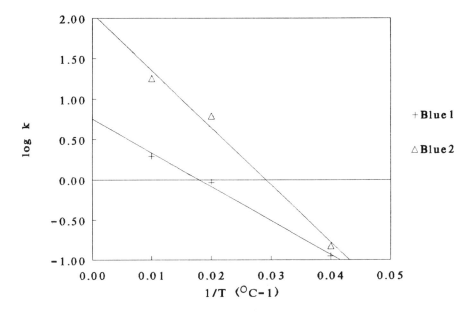

Figure 5.   Arrenhius plot of temperature dependence
of rate constants of dye bath depletion by wool
exposed to 20 ppm dye solutions.

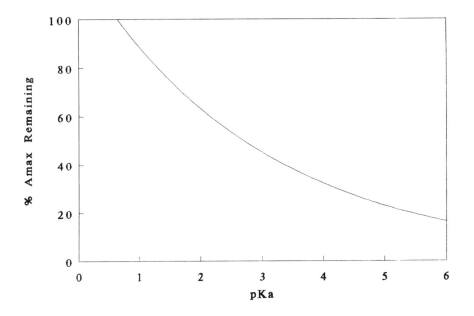

Figure 6.   Changes in dye bath depletion with pKa
variations.   Nylon 6.6 exposed for 1 hour to 20 ppm
Blue 1 at 25°C.

From these results, we have established that the mechanism of the dye to amine end interaction is one of charge to charge interaction between the dye anion and the cationic functional group on the substrate to form an electrostatic bond. At high acid content, the amine site on the substrate is ionized (6). In the case of proteins, salt linkages between chains are broken due to protonization of carboxyl ends leaving the positively charged amine ends available for dye binding.

The values obtained from studying the dye uptake by nylon and wool as a function of time were used to calculate the rate constants. A plot of the decrease in dye present in the bath as a function of time typically showed an initial linear portion until the dye was exhausted and the curve levelled off. This was evidence that the staining followed first order kinetics. Linear regression fits of the linear portions of the curves produced slope values which were the first order reaction rate constants at 100°C (Table V).

It was found that proteins depleted dye baths faster than equivalent amounts of nylon. This was a consequence of faster diffusion of dye due to the higher moisture content of proteins as well as proteins' more irregular surface being more conductive to the retention of hydrophyllic molecules (7). This faster attachment does not mean protein bound dye was more tightly held. It was found through some preliminary studies, that migration of the dye occurred between proteins and nylon indicating nylon holds the dye molecules tighter. Diffusion occurred from wool to nylon but not from nylon to wool. This means that colored food proteins are capable of staining nylon packages but that color bleed from the package into the food is unlikely.

TABLE V. k Values (Min-1)

Color	Nylon	Wool
Blue 1	0.878	1.97
Red 40	2.72	3.62
Yellow 5	1.59	3.02

The number of sulfonic acid groups present on the dye had an effect on depletion rates. The calculated rate constants proved that the colors containing two acid groups, Red 40 and Yellow 5, were depleted from the dye bath faster than the Blue 1 which has three (Table V).

Additive Effects. In the interest of possibly manipulating dye migration from food to other polymers

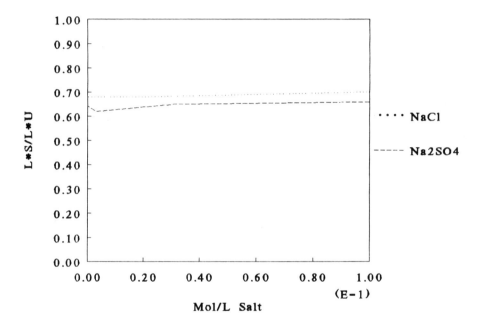

Figure 7.   Changes in substrate staining with
increasing salt concentration.   Nylon 6.6 exposed to
20 ppm Red 40 for 1 hour at 50°C.

by using additives, the effects of some common salts and sugars on dye to polymer interactions were also investigated. No enhancement or decrease in the intensity of the stain was found by salt addition. Figure 7 is a typical curve for these trials.

Sugars are common additives to foods containing the synthetic food dyes. Sucrose and its component monosaccharides, fructose and glucose were not found to have any effect on enhancing or reducing the intensity of the stain.

CONCLUSION

In summary, the interaction between the F D and C dye molecules and polymers containing protonable amine end groups is essentially an acid-base reaction. The depletion of colors from food materials involves the amine end groups on food and nonfood polymers. The results indicated that the greatest amount of coloring was achieved at a pH value of 3.0. By adjusting food formulations to lower pH values, the greatest color yield is obtained, but the conditions are also set for greatest dye migration from food to packages. Packaging designs have to take into consideration the presence of protonatable amine groups on the polymers if they are to come into contact with foodstuffs containing the F D and C colors in the dye form.

If liquids or moist foods with high water contents are in direct contact with the polymer, the interactions of food dyes with packaging materials are of primary interest from an aesthetic standpoint. Staining of the package and loss of color of the contents have been observed in actual applications. Some examples are prepared gelatin desserts packaged in plastic tubs and colored beverage powders in plastic lined envelopes. This work has included only the 6 type nylons and no work has been done with the newer aromatic materials that have less sensitivity to water.

Acknowledgments. The authors wish to thank Ms. Katy J. Dishart and Mr. Edward W. Fassler for technical assistance and Mrs. Yoga Pandya for making the figures.

Literature Cited

1. Aurand, L.W.; Woods, A.E.; Wells, M.R. Food Composition and Analysis; Van Nostrand Reinhold: New York, N.Y., 1987; p. 471.
2. Briston, J.H. Plastic Films; Longman Scientific and Technical Group, Ltd.: London, U.K., 1989, p. 350.
3. Johnson, F. Private Communication. E. I. Dupont de Nemours and Company, Inc., 1989.

4. Remington, W. R.; Gladding, E. K. <u>J.</u> <u>Amer.</u> <u>Chem.</u>
    <u>Soc.</u> **1950,** 72, pp. 2553-2559.
5. Vickerstaff, T. <u>The</u> <u>Physical</u> <u>Chemistry</u> <u>of</u> <u>Dyeing;</u>
    Interscience Publishers: New York, N.Y., **1954,**
    pp. 346, 157, 462.
6. Rush, J.L. <u>J.</u> <u>Text.</u> <u>Chem.</u> <u>Colorists,</u> **1980,** 12(2),
    pp. 35/25 - 37/27.
7. Sadov, F.; Korchagin, M.; Matetsky, A. <u>Chemical</u>
    <u>Technology</u> <u>of</u> <u>Fiberous</u> <u>Materials;</u> Mir Publishers:
    Moscow, U.S.S.R., **1973,** p. 111.

RECEIVED March 1, 1991

Chapter 5

Application of a Poly(tetrafluoroethylene) Single-Sided Migration Cell for Measuring Migration through Microwave Susceptor Films

Timothy H. Begley and Henry C. Hollifield

Indirect Additives Section, Division of Food Chemistry and Technology, Center for Food Safety and Applied Nutrition, Food and Drug Administration, 200 C Street, S.W., Washington, DC 20204

Conducting migration studies on microwave susceptor packaging presents a challenge because of the high temperatures attained during microwave cooking. Neither conventional metal migration cells nor typical food-simulating solvents can be used. This paper describes the use of a polytetrafluoroethylene single-sided migration cell for monitoring the movement of migrants through polymeric films during microwave heating. Data are presented on the permeation of plasticizers through polyethylene terephthalate, polyether imide, polyethylene naphthalate, and polycyclohexylenedimethylene terephthalate susceptor films.

Advances in food packaging technology during the past few years have greatly increased the temperature range over which packaging materials are used. In particular, microwave susceptor packaging is exposed to temperatures from below 0 °C to over 220 °C. For the higher temperatures, little or no information is available on the barrier properties of the food contact layer, which is usually polyethylene terephthalate (PET). Recently it has been demonstrated that some plasticizers penetrate the PET susceptor films and migrate into the food (1).

The gas barrier and permeation characteristics of PET are generally well known. Typically the permeation experiments are performed in closed systems at 25 °C with a duration of many hours. These test conditions are not applicable to susceptor films because under actual use conditions the films heat from below 0 °C to over 220 °C in less than 5 min. This paper describes the use of a polytetrafluoroethylene (Teflon) single-sided migration cell for evaluating the permeabilities of susceptor films of different polymer construction under actual use conditions. In addition, the cell is used to demonstrate the relative amounts of residual polymer components of low molecular weight that migrate under susceptor heating conditions. The films used were PET, polyether imide (PEI), polyethylene naphthalate (PEN), and polycyclohexylenedimethylene terephthalate (PCDMT).

Experimental

In the tests a piece of fortified paperboard was covered with a susceptor film. The covered paperboard was placed in a single-sided migration cell, a food-simulating oil was added, and the cell was microwaved. Then the oil was extracted to determine the penetrants and migrants. The oil extract was analyzed by high-performance liquid chromatography (HPLC).

Susceptor Films. The PEI susceptor film, supplied by General Electric Plastics, Pittsfield, MA, was made from ULTEM[R] resin and had a nominal thickness of 0.0254 mm. The PCDMT and PEN susceptor films were supplied by Eastman Chemical Products, Kingsport, TN. PCDMT was produced from the KODAR THERMX[R] resin and was nominally 0.0254 mm thick; PEN was an experimental film and was nominally 0.0381 mm thick. The PET susceptor film, supplied by Waldorf Corporation, St. Paul, MN, was 0.0127 mm thick and was the only film which was biaxially oriented.

Residual UV-absorbing low molecular weight polymer components in the susceptor films were determined by a precipitation method described by Begley and Hollifield (2). The method was modified slightly for PEI by using methylene chloride to dissolve the polymer and methanol to precipitate the bulk polymer.

Paperboard Preparation. A circular paperboard disk, 3.5 inches (8.9 cm) in diameter, was cut from virgin paperboard flat stock 0.018 inches (0.457 mm) thick, similar to paperboard used for microwave susceptor construction. The paperboard disk was placed in a glass petri dish and treated with 1.0 mL of a standard solution containing 20 mg/mL each of diethylene glycol dibenzoate (DEGDB), benzaldehyde (BA), butyl benzyl phthalate (BBP), and 2-butoxyethanol (2BE) in methylene chloride. The petri dish was covered for 5 min; then the cover was removed for 3 min to let the solvent evaporate. The paperboard was transferred to the microwave cell, covered with a circular piece of susceptor film 3.5 inches (8.9 cm) in diameter, and held in place with the seal ring.

Microwave Migration Cell. An illustration of the microwave cell used in this work is given in Figure 1. The original cell was designed at Waldorf Corporation, St. Paul, MN. We have made minor modifications to the cell to suit our experimental needs; for permeability studies, for example, a vent hole **must** be added in the bottom base plate. Without the vent hole the susceptor film will be forced off the paper surface by vapors generated during the heating process, and abnormally low permeabilities will result.

Then 30.0 g of Miglyol (Miglyol 812, fractionated coconut oil with a fatty acid composition of 50% C_8 and 45% C_{10}) was added, and the cell was microwaved for 4 min in a 700-W Amana Radarange microwave oven. Under these test conditions, no cracking of the test susceptor films was observed.

The 30 g of Miglyol in the cell corresponds to a ratio of mass of oil to susceptor surface area of 4.2 g/in^2. In typical commercial packages the ratios of food mass to susceptor surface area range from 2.3 to 4.2 g/in^2. The microwave oven had an output wattage of 719 W (measured by heating 1000 g of water and measuring the net temperature change). The temperature of the Miglyol during microwaving was measured with a Luxtron Model 755 fluoroptic temperature-sensing instrument (Mountain View, CA) interfaced to an IBM-AT computer. When microwaving was completed, a 5-mL gas-tight syringe was used to withdraw a 3.0-g test portion of the oil from the cell for extraction and determination of penetrants and migrants.

Extraction of Penetrants and Migrants. In this paper penetrants are defined as those chemical species, including adhesive components, that leave the paperboard and diffuse through the susceptor film and into the oil. Migrants are defined as those chemical species that originate in the susceptor film and diffuse through the susceptor film into the oil.

The penetrants from the fortified paperboard and migrants from the test film were extracted from the oil by using a hexane/acetonitrile partitioning procedure described in the method for the determination of PET oligomers in corn oil (3). Because 2BE does not absorb UV light, the method was modified slightly to incorporate the preparation of a 2BE derivative that absorbs UV light. The derivatization procedure simply involved adding dinitrobenzoyl chloride, along with dimethyl acetamide as a catalyst, to the acetonitrile extract in a Kuderna-Danish concentrator. The 2BE dinitrobenzoyl derivative was formed during solvent removal. The acetonitrile was then concentrated to 5 mL, and penetrants and migrants were determined by HPLC. All quantitative analyses used external calibrations.

HPLC Standards. To estimate the total migration of low molecular weight residual polymer components, model compounds whose structures resembled those of the monomer units were used as HPLC quantitative standards. Area response factors were determined for the model compounds and used to estimate the relative migration from the various polymers. The PET cyclic trimer was used to estimate migration from PET and PCDMT. Phthalimido-acetaldehyde diethylacetal was used to estimate migration from PEI, and 2,6-naphthalene dicarboxylic acid was used to estimate migration from PEN.

HPLC System. The HPLC system consisted of a Hewlett-Packard Model 1090 liquid chromatograph equipped with a 20-μl injector loop, an additional Rheodyne Model 7125 injector valve with a Brownlee 5-μm C_8 guard column in place of the injector loop, a 5-μm Microsorb C_8 250 X 4.6 mm column, a Waters Model 480 Lambda Max variable wavelength detector operated at 240 nm, and a Nelson Analytical Model 3000 chromatography data system operating on an IBM AT computer. The HPLC mobile phase was as follows: solvent A, water:acetonitrile (85:15); solvent B, 100% acetonitrile. A linear solvent gradient was programmed at a flow rate of 1.5 ml/min as follows: from 30% B to 100% B in 20 min; 100% B for 4 min; from 100% B to 30% B in 1 min.

Results

Representative heating profiles for all the susceptor films are illustrated in Figure 2. The heating curve for cooking a frozen pizza in a commercial susceptor package is included to demonstrate the similarity of the migration test cell conditions to actual cooking. All the susceptor films have different heating rates, as also shown in Figure 2. Each film achieves a different maximum temperature, with PEI achieving the highest (240°C) and PCDMT the lowest (205°C). All films achieve temperatures greater than their glass transition temperatures (T_g) under the test conditions and thus would be expected to be permeable to adhesive and/or paperboard components.

Because of the temperature variability of the Miglyol as a result of being heated by different susceptor films (Figure 2), control experiments were performed to determine the effect of these temperature differences on the amount of each penetrant present in the Miglyol. It is assumed that the differences in penetrant concentration in Miglyol after it is heated by the different susceptor films are mostly due to volatilization of the penetrants. Temperature control factors were determined by calculating the ratio of penetrant concentration in unheated fortified Miglyol to that in susceptor-heated fortified Miglyol. The Miglyol was fortified with 20 μg/g each of BA, DEGDB, and BBP.

The temperature control factors for each penetrant and each polymer susceptor film are listed in Table I. BA, because of its large control factor, is readily volatilized from the hot Miglyol (see Table I). Because of its high volatility under the test conditions, BA can be considered only as a qualitative indicator of the

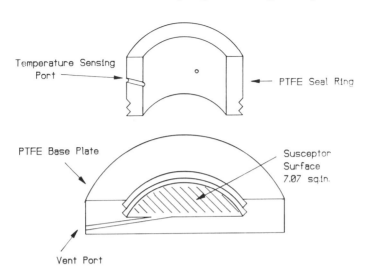

Figure 1. Drawing of microwave migration cell.   The cell is a two-piece design made of polytetrafluoroethylene. The actual cell used contained three vent ports and three temperature-sensing ports.

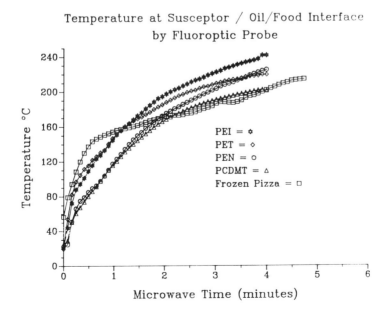

Figure 2.   Temperature versus time heating curves for the different susceptor films and a frozen pizza.

susceptor film's barrier properties towards aromatic
aldehydes. The control factors listed in Table I show that
DEGDB and BBP are more firmly retained in the Miglyol
during microwave heating than is BA. In particular,
because all the BBP control factors are within
approximately 3% of unity (3% is the relative variability
of extracting BBP and DEGDB from Miglyol), BBP is
quantitatively retained within the migration cell under
all four susceptor heating conditions. Susceptor heating
of Miglyol causes DEGDB losses from Miglyol of 5% to 21%
depending on which susceptor film is used to heat the
Miglyol (Table I).

Table I. Temperature Normalization Factors for Heating
Fortified Miglyol[a]

Susceptor Film	BA	DEGDB	BBP
PET	6.60	1.21	1.03
PCDMT	4.94	1.05	0.97
PEN	5.36	1.08	0.99
PEI	20.2	1.14	1.02

[a]Each value represents the ratio of the concentration in
unheated Miglyol to the concentration in Miglyol heated
by the susceptor.

Table II. Permeability of Susceptor Films

Susceptor Film	ng/cm^2s		
	BA	DEGDB	BBP
PET	71.3[a] ± 23.8[b]	190 ± 48	74.4 ± 21.9
PCDMT	151 ± 21	170 ± 61	85.6 ± 38.8
PEN	7.9 ± 11	5.5 ± 4.4	3.0 ± 2.2
Commercial susceptor		107 ± 4.1	

[a]Each value is the mean permeation of four susceptor films
of a given type.
[b]Each ± value is the standard deviation of four permeation
values.

Permeability of Susceptor Films. Typical chromatograms
for the determination of the penetrants in acetonitrile
extracts of Miglyol are illustrated in Figure 3, which
shows that these penetrants are easily separated and
identified in the Miglyol extracts.
    The results for the permeability tests on the
susceptor films are listed in Table II. All the values
reported in Table II have been adjusted by using the

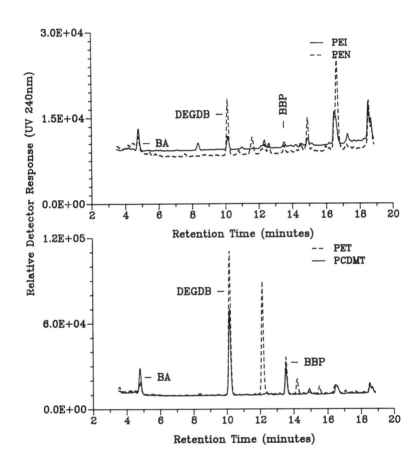

Figure 3. Chromatograms obtained from analysis of Miglyol for penetrants of the various susceptor films.

temperature control factors listed in Table I and
normalized for differences in susceptor film thickness.
No adjustments have been made to the data in Table II for
the amount of metalizing, the degree of crystallization,
or the degree of orientation of the films, all of which
affect the permeable to some degree, with PEI being the
least permeable and PET and PCDMT the most permeable.

Although the data generally show the films to be a
better barrier to BA than to DEGDB and BBP, this
interpretation may not be justified. The films are
probably just as permeable to BA as to DEGDB and BBP, but
because of large losses of BA due to volatilization, as
indicated by the large temperature control factor in Table
I, the data do not accurately reflect the permeability of
the films to BA compared to DEGDB and BBP. In a
qualitative sense, however, the data do show that PEN is
a better functional barrier for BA than are PET and PCDMT
because much more BA penetrates PET and PCDMT than PEN,
whereas PET, PCDMT, and PEN all cause approximately the
same amount of loss of BA from Miglyol.

The permeation data from Table II for DEGDB and BBP
suggest that there are distinct differences in the barrier
properties of the susceptor films. Included in Table II
is the permeation of DEGDB from a commercial susceptor to
show that our simulated susceptors (a susceptor film
placed on top of paperboard and held in place by the seal
ring to the cell) give reasonable values for the given
test conditions. As these data clearly show, the PEI
susceptor film has the best barrier properties of all
those tested (at least 10-fold greater than commercial PET
susceptor film). This is expected because PEI has a very
high T_g (213°C) compared to the other susceptor films (PET
T_g = 80°C, PCDMT T_g = 94°C, PEN T_g = 120°C). Below the T_g,
diffusion through polymers in a glassy state is very slow
and therefore the permeability of the film will be small.
Low permeabilities would be expected because the
temperature of the PEI/Miglyol interface under the given
test conditions is below its T_g for 50% of the test time
(see Figure 2). The PEN susceptor film generally shows
twice the barrier properties of the commercial PET film,
while the PCDMT film appears to have barrier properties
equivalent to those of PET. The barrier properties of PEN
and PCDMT may be further improved if the films are
biaxially oriented.

The data in Table II also clearly show that at least
twice as much DEGDB as BBP penetrates the susceptor films.
This finding is consistent with recent work by Mauritz et
al. (4-6) who, using free volume-based theory, have shown
that less diffusion occurs above the polymer T_g for
geometrically broad (fat) molecules compared to
geometrically elongated (thin) molecules of comparable
molecular masses (molecular weight is 312 for BBP and 314
for DEGDB). Because BBP could be considered a fat
molecule relative to DEGDB, it should diffuse more slowly.

Assuming that DEGDB and BBP have similar solubilities in the films (a reasonable assumption because the calculated Hildebrand solubility coefficients are 10.0 $[cal/cm^3]^{\frac{1}{2}}$ and 9.9 $[cal/cm^3]^{\frac{1}{2}}$ for DEGDB and BBP, respectively (7)), BBP will have a lower permeation, as experimentally observed.

Although attempts were made to quantitate the amount of 2BE that penetrated the films, no reliable data were obtained. It was expected that 2BE would penetrate the films to approximately the same extent as the plasticizers, or possibly with some losses due to volatilization and higher adsorption to the paperboard. However, this was not the case; essentially no penetration was observed. Investigations using fortified paraffin oil and fortified Miglyol indicated that 2BE reacts with the Miglyol under susceptor heating conditions. It is also possible that some 2BE could react with PET, PCDMT, and PEN at the test temperatures because the polymerization catalyst is still present in the films. Further work is planned to investigate the permeation characteristics of 2BE.

To further explore the permeation characteristics of the susceptor films, the permeability values for DEGDB and BBP were plotted versus T_g (Figure 4), demonstrating what appears to be a linear relationship between the permeability of the susceptor films and the T_g of the films. Therefore, the higher the T_g of the susceptor film the greater the barrier properties of the film under microwave susceptor heating conditions.

Because the plasticization process is known to lower the T_g of a polymer, thus affecting its permeation characteristics, control experiments were performed in which residual polymer migration was compared for films heated in contact with fortified and unfortified paperboard. If the films became sufficiently plasticized to alter the diffusion process in the film under the test conditions, an increase in residual polymer component migration would be observed. Within the precision of the test, no significant increase in residual polymer component migration was observed; therefore the presence of the plasticizers in the film at the levels encountered under the test conditions did not affect diffusion in the film.

Residual Polymer Component Migration. Concurrent with the permeation tests, migration of residual low molecular weight polymer components can be evaluated. Initially the residual polymer components must be identified so that appropriate chromatographic conditions can be established for determination of the components that migrated to Miglyol. The residual polymer profiles have been generated by the procedure cited above and are illustrated in Figure 5. The relative UV responses in the chromatograms have been

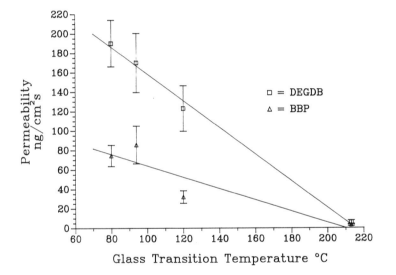

Figure 4. Plot of permeability of microwave susceptor films versus glass transition temperature T_g during a 4-min microwave heating. The error bars represent the standard deviations listed in Table 2.

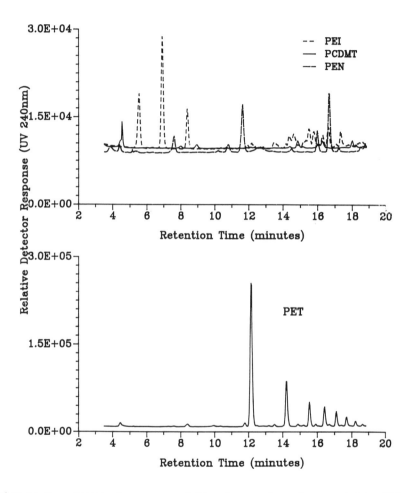

Figure 5. Chromatograms of the low molecular weight, UV-absorbing residual polymeric components in the susceptor films.

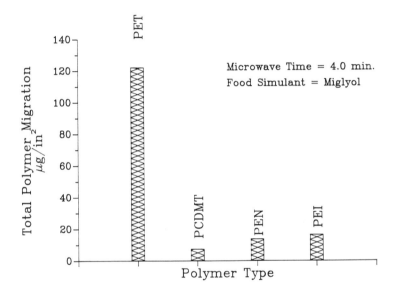

Figure 6. Histogram of the total migrating low molecular weight, UV-absorbing residual polymeric components from susceptor films during a 4-min microwave heating.

normalized to give a relatively quantitative picture of the amounts of residual low molecular weight polymer components in the susceptor films. All of the tested susceptor films contain residual low molecular weight polymers (Figure 5). The extent to which these components migrate into Miglyol under susceptor heating conditions is shown in Figure 6. This histogram gives the estimated total migration of polymer residuals into Miglyol after 4-min heating and shows that the migration from a PET susceptor film is 10-fold greater than from the other susceptor films. This is expected because, as Figure 5 shows, PET has at least 10-fold greater residual low molecular weight polymer components than the other susceptor films. Unlike the data listed in Table II, the data in Figure 6 have not been adjusted for differences in temperature. The lack of temperature adjustment is not expected to significantly alter the overall picture; that is, much more migrates from a PET susceptor film than from PCDMT, PEN, and PEI films because each of the latter three has much lower quantities of polymer residuals than PET, as illustrated in Figure 5.

The results for migration of residual low molecular weight polymers and permeabilities of the susceptor films lead to the following general conclusions. All the susceptor films were permeable to BA, DEGDB, and BBP. Penetration occurred in the absence of cracking or crazing of the susceptor films, which is normally seen when food is cooked by a commercial susceptor. This study demonstrates that PET susceptor films **do not** prevent adhesive components from migrating to fatty food and that the total migration of low molecular weight residual polymer components through PET is 10 times that of PCDMT, PEN, and PEI susceptor films. Although the permeation and migration characteristics of the films held in the PTFE cell are not necessarily characteristic of actual susceptor constructions because of lack of adhesion, the data presented here illustrate that relative permeation and migration characteristics can be measured using the PTFE cell.

Acknowledgments

We thank the following companies for supplying materials for this study: Eastman Chemical Products, Kingsport TN; General Electric Plastics, Mt. Vernon, IN; and Waldorf Corporation, St. Paul, MN.

Literature Cited

1. Begley, T.H.; Hollifield, H.C. J. Food Prot. **1990**, In press.
2. Begley, T.H.; Hollifield, H.C. J. Assoc. Off. Anal. Chem. **1989**, _73_, 468-470.
3. Begley, T.H.; Hollifield, H.C. J. Agric. Food Chem. **1990**, _38_, 145-148.

4. Mauritz, K.A.; Storey, R.F.; George, S.E.
   Macromolecules, **1990**, <u>23</u>, 441–450.
5. Mauritz, K.A.; Storey, R.F. <u>Macromolecules</u>, **1990**, <u>23</u>,
   2033–2038.
6. Coughlin C.S.; Mauritz, K.A.; Storey, R.F.
   <u>Macromolecules</u>, **1990**, <u>23</u>, 3187–3192.
7. <u>Polymer Handbook</u>, 2nd Edition; John Wiley & Sons, New
   York, 1975.

RECEIVED March 1, 1991

Chapter 6

Determining Volatile Extractives from Microwave Susceptor Food Packaging

William P. Rose

The National Food Processors Association, Western Research Laboratory, 6363 Clark Avenue, Dublin, CA 94568

Under the auspices of a coalition of The National Food Processors Association members and members of the Society of Plastic Industries a method for the determination of volatile extractives from microwave susceptor containing food packaging was developed. The method consists of a headspace sampling technique followed by Gas Chromatography with Flame Ionization or Mass Spectroscopy detection. A portion of the package containing the microwave susceptor is subjected to microwave cooking under simulated use conditions. An aliquot of the headspace gases are then introduced into a gas chromatograph for the determination and identification of individual compounds. 4-Heptanone was used as an internal standard. Susceptor packages representing a popcorn bag and a Pizza package were fortified with eight compounds representing a range of polarities. Recoveries of these compounds ranged from 22 to 98 percent with coefficients of variation of 3 to 24 percent. Compounds identified by Gas Chromatography / Mass Spectroscopy in Microwave Susceptor Packaging representative of the packaging found in the marketplace ranged in molecular weight from 32 to 200. The boiling points of these compounds ranged from 21 to 216°C.

The use of Microwave Susceptor Technology for food packaging is a rather recent development. The development of this food packaging technology was a direct outgrowth of the technology developed for the stealth bomber. When applied to the stealth bomber this technology enabled the bomber to absorb radar radiations and thus escape detection. In the food packaging applications that we are most

0097–6156/91/0473–0067$06.00/0

familiar with, the microwave susceptor consists of a metalized PET (Polyethylene Terphthalate) film laminated with an adhesive to one or more layers of paper or paper board stock. The metalized film absorbs the microwave radiation producing a rapid, localized, high temperature build-up. These heat susceptors are used to crisp or brown the food product which does not usually crisp or brown under conventional microwave cooking conditions. Some of the typical uses for microwave susceptors include the crisping or browning of pizza, popcorn, batter fried sea food, and pastries. More and more applications for microwave susceptors are appearing in the marketplace each month. This leads us into the need for microwave package testing and regulation.

The FDA first became aware of the possibility of microwave packaging component migration from microwave susceptors during the winter of 1987-1988. Some of the FDA scientists had related their home experiences with microwave popcorn bags that developed burn holes and burnt product during cooking. The FDA scientists discovered that the temperatures of the microwave susceptors in these packages sometimes reached temperatures in excess of 500°F.

All of the currently used packaging materials are approved for food contact use under existing FDA regulations. However, these regulations and tests were developed in 1958 and only considered reactions occurring at temperature up to 300°F. The FDA scientists conducted some preliminary experiments with microwave susceptors and found that some of the components of the microwave susceptor package did indeed migrate from the package into a food simulant. In October of 1987 an internal memo was circulated in FDA expressing concern about "active" microwave packaging. In January 1988 Breder's paper further aroused the interest of the press and industry.

In August of 1988 The National Food Processors Association and The Society of Plastic Industries joined forces to form the Microwave Susceptor Packaging Committee. The committee consisted of representatives of the manufacturers of food, packaging materials, and adhesives. Chemists from the FDA also sat in on the meetings. In September 1988 the FDA requested that manufacturers of microwave food and packaging materials supply a wide variety of chemical and toxilogical information on the substances used to manufacture microwave packages. A year later the FDA published an announcement of proposed rule making in the Federal Register requiring the submission of microwave packaging data. This work represents part of the Microwave Susceptor Committee's response to FDA's request and requirements. It is important to remember that FDA did not suggest that there was a question regarding the safety of susceptor packaging. FDA based it's concerns on the lack of data on actual use temperatures and the migration levels at these temperatures.

Selection of A Method for Determining Volatile Extractives

The selection of a method for the determination of volatile extractives from microwave susceptor packaging had to meet three important criteria: (1) The method would have to produce results that would adequately represent those found under actual use conditions; (2) The method had to have satisfactory precision and accuracy; and (3) The method had to be sensitive enough to provide data which could be used to assure the safety of susceptors, and also be used as a screening tool.

Method Sensitivity. In establishing the required sensitivity for any analytical method, it is necessary to consider the purpose for which the data is being obtained. In the case for susceptors, the primary purpose is to demonstrate that there are no health or safety concerns for the continued use of these packaging materials. The Susceptor Microwave Packaging Committee, taking into consideration the estimated daily intake, the consumption factor for the food product packaged in the susceptor package, and using the FDA's ratio of weight of food/square inch of food package surface area of 10 g/in^2, established the required method sensitivity of 10 $\mu g/in^2$ of susceptor. However, to adequately asses and assure the safety of susceptors, the committee decided not to identify and determine volatile substances in the <u>diet</u> at this sensitivity, but to identify and determine them at the same sensitivity in the <u>contacted food</u>. This resulted in a required sensitivity of 0.5 $\mu g/in^2$ of susceptor surface. In view of the fact that the committee used highly conservative estimates, and applied them only to substances not known to be animal or human carcinogens, any substance observed in the test vial headspace at concentrations of less than 0.5 $\mu g/in^2$ need not be considered to raise health or safety concerns. A method sensitivity of 0.5 $\mu g/in^2$ would yield a conservative calculated concentration in the diet of 0.5 ppb. A more detailed discussion of the factors involved is contained in the report of The Susceptor Microwave Susceptor Committee to the FDA (*1*).

Method Selection. No standardized methodology for the determination of volatile extractives was in place at the time this work was initiated. A number of methodologies were being used at the time in the laboratories of the committee members. Some of these methods were suggested for use in this work. The suggested methods were examined closely for their applicability to the purposes of the committee. In general the suggested methods fell into three types; (1) purge and trap, (2) diffusion trapping, and (3) static headspace sampling.

Although purge and trap systems theoretically offer lower detection limits than static headspace techniques, they are difficult systems to install and maintain in a microwave oven. Further experience and calculation showed that the increased sensitivity was not really necessary.

Diffusion trapping is not a commonly used technique compared to the well known and used static headspace sampling technique. In addition, no written methodology was available for the diffusion trapping technique which could be considered by the committee. Also the technique, as was understood by the committee, had a number of questionable manipulative operations which raised concerns about quantitation, the applicability to actual use conditions, and the time required to complete an analysis.

Additional data was collected for the static headspace and the other suggested methods through a round robin study using the same microwave susceptor sample for all the methods examined. After reviewing the various methods with particular attention given to reproducibility, sensitivity, and sample heating conditions simulating actual consumer use, a method using microwave heating, static headspace sampling, and GC analysis employing flame ionization detection (FID) was selected as having the best potential for meeting the committees requirements.

Methodological Details

Several goals and assumptions were made in the development of this method. The method would have to simulate actual use conditions. Using the food/susceptor system itself would produce to many volatiles from the food product which could not be easily separated from the susceptor volatiles. The temperatures reached by the susceptor during actual use will be highly dependant upon the food system being used. Therefore the susceptor sample would have to undergo microwave heating in such a manner that the time/temperature profile of the susceptor interface would simulate that of the susceptor when used with the actual food load. This was accomplished by obtaining the time/temperature profile of the susceptor interface while microwaving the actual food product. The measurements were made in a calibrated (700 +/-15 watt) microwave oven using a Luxtron Model 755 Fluroptic Thermometry System (Mountain View, CA). Temperature readings were recorded every 15 seconds during the cooking cycle recommended in the package instructions.

A one square inch sample of the susceptor was then placed in a 10 ml reaction vial sealed with a teflon faced septum crimp seal. The septum was pierced and a Thermoptic probe inserted through the septum in such a manner that the tip rested at as acute an angle as possible in the center of the susceptor sample. Various water loads were then placed in designated positions in the oven until a time/temperature profile as close as possible to the actual use time/temperature profile was obtained. This was done for each type of susceptor/food product tested. An example of one such time/temperature plot is shown in figure 1. Samples were then prepared in an identical manner except that the Luxtron probe was omitted and a known amount of a water solution of 4-Heptanone (usually 10 ul) was added as an internal standard. After microwaving the samples were allowed to come to equilibrium in a 90°C convection oven for three minutes. Using a heated syringe equipped with a valve, a two ml sample of the headspace gases were withdrawn from the vial and injected into the Gas Chromatograph for analysis. The Gas Chromatographic conditions are given in Table I. The chromatographic conditions are identical for both GC/FID and GC/MS analyses.

No attempt was made to determine the partition coefficients between the food system and the headspace gases. It was assumed that 100% of the gases would migrate into the food system. Thus the results of these analyses can be considered as being an extreme of a worst case scenario.

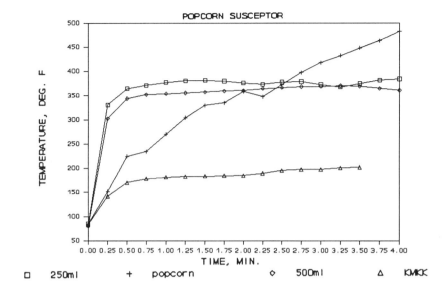

Figure 1. Microwave Heating Time/Temperature.

Table I. Gas Chromatographic Conditions

Instrument: Hewlett/Packard 5880A
Carrier Gas: Helium
Column: DB-5, 0.23mm x 30m x 1um
Carrier Flow-Approx: 1 ml/min. (20-25 cm/sec)
Temperature program:
    Splitless injection: 15 seconds
    Initial Temperature: 30°C
    Initial hold: 5 minutes
    Program rate: 4°C/minute
    Final temperature: 240°C
    Final hold: 10 minutes
Gas flow rate for FID:
    Hydrogen: 25 ml/minute
    Air: 350 ml/minute
    Makeup gas: 24 ml/minute(Total flow 25 ml/ minute)

Method Validation

Seven compounds were selected for use in the validation of the method. These compounds were selected to give a range of compound types and polarities. They were also selected as compounds that were likely to be encountered in the volatile extractables. In fact, some of these compounds were already known to be present in the volatile extractives. The compounds selected for use in the validation of the method included Acetone, Furan, Butanol, Benzene, Toluene, Furfuraldehyde, and decane. Standard solutions of these compounds were prepared at various concentrations in methanol. Solutions of these compounds prepared in water proved to be unstable and not all of the compounds were miscible with water at the desired concentrations.

Response Factors and Standard Curves. Ten microliters (ul) of the standard solutions were placed in an empty 10 ml reaction vials crimp sealed with teflon faced silicon septum seals. The vials and standard solutions were allowed to come to equilibrium at room temperature for one hour. The reaction vial was then heated in a convection oven at 90°C for three minutes. Two mls of the headspace gases were then withdrawn from the reaction vial with a valved heated syringe and injected into the gas chromatograph for quantitative analyses. Results of the Least Squares Curve Fit on each of the compounds are shown in Table II.

Table II. Standard Curves, Reaction Vial without Susceptor, Treated Same as Samples, Nominal 5, 10, and 20 $\mu g/in^2$, Triplicate Analyses. Least Squares Curve Fit, Y = A + (B * X)

Compound	A	B	Correlation Coefficient	Coefficient of Variation, %
Acetone	-0.387	2.300E-05	0.9975	4.49
Furan	4.743	2.552E-05	0.9874	2.05
Butanol	-0.310	2.215E-05	0.9979	4.20
Benzene	0.496	1.842E-05	0.9992	2.29
Toluene	-0.596	1.772E-05	0.9990	2.05
Furfural	-0.122	4.431E-05	0.9952	4.01
Decane	-0.294	2.097E-05	0.9997	1.34

A second series of standards were also run at a low concentration ranging nominally from 0.1 to 1 $\mu g/in^2$. This data is presented in Table III for comparison. This second series of standard solutions was prepared in water and may account for the slight differences found between the two sets of data. Also the differences reflect the lower concentration and thus a wider coefficient of variation.

Recovery Studies with Selected Compounds. Recovery of selected compounds from a microwave susceptor popcorn bag and pizza board were conducted using the same series of compounds as the standard curves. In this case one square inch of the susceptor sample was placed in the reaction vial. Ten microliters of the mixed standard solution was placed in the middle of the susceptor surface. The vial was sealed and

Table III. Standard Curves, Reaction Vial without Susceptor, Treated Same as Samples, Nominal 0.1, 0.5, and 1 $\mu g/in^2$, Triplicate Analyses. Least Squares Curve Fit, Y = A + (B * X)

Compound	A	B	Correlation Coefficient	Coefficient of Variation, %
Acetone	-0.068	2.242E-04	0.9821	4.82
Furan	0.117	7.790E-05	0.9773	3.74
Butanol	0.128	6.155E-05	0.9681	3.16
Benzene	0.111	2.064E-05	0.9926	8.17
Toluene	0.304	3.065E-05	0.7848	50.83
Furfural	0.054	7.448E-05	0.9922	8.06
Decane	0.157	5.036E-05	0.9158	4.27

Table IV. Recovery Studies from Popcorn Bag, Nominal 5, 10, and 20 $\mu g/in^2$, Triplicate Determinations

Compound	Average % Recovery	Coefficient of Variation
4-Heptanone	68.72	8.16
Acetone	98.56	7.16
Furan	82.16	12.65
Butanol	86.49	6.88
Benzene	60.59	6.12
Toluene	92.12	3.95
Furfuraldehyde	21.88	16.41
Decane	95.28	3.25

Table V. Recovery Studies From Pizza Board, Nominal 5,10, and 20 $\mu g/in^2$, Triplicate Determinations

Compound	Average % Recovery	Coefficient of Variation
4-Heptanone	77.40	8.99
Acetone	86.58	12.51
Furan	80.44	23.87
Butanol	77.91	9.46
Benzene	67.08	8.96
Toluene	88.81	3.46
Furfuraldehyde	30.95	16.87
Decane	83.58	3.94

microwaved according to the package instructions. For popcorn it was found that the optimum cooking time in a 700 watt oven on the high setting was 3 minutes 15 seconds. This was the time used for the microwave heating of the sample.

Results of the recovery studies for both types of susceptors are shown in Tables IV and V respectively. Recoveries ranged from 60% for benzene to 98%

for acetone in the popcorn bag sample. The results for the pizza board are similar with recoveries of 67% for benzene and 86% for acetone. The exception to these recoveries was furfuraldehyde. Recoveries for furfuraldehyde were 22 and 31 %.

Sample Analyses

Susceptor samples for testing were requested by food company members of the committee from their suppliers. The samples were submitted through Keller and Heckman, Counsel to the Susceptor Microwave Packaging Committee. The suppliers were asked to submit the samples in the form in which they were intended to be used. In other words they should represent the complete susceptor construction containing the microwave susceptor construction itself, the adhesives, paper, and in some cases the printing inks. The samples were received in this laboratory wrapped in a double layer of aluminum foil to prevent external contamination and/or possible loss of volatiles. Each sample had been identified by a unique code that was known only to Keller and Heckman. The names of the submitters and the sample codes were maintained as confidential information by Keller and Heckman.

In all, 42 different samples were received from 15 different companies for testing. The samples received for testing are summarized in Table VI.

Table VI. Samples Analyzed

Popcorn(13)
Pizza(9)
Fish(7)
Brownies(3)
Pot Pies(3)
Other(7)

Analyses by GC/FID. A total of 44 commercially available and developmental microwave susceptor samples were analyzed by the validated method described above. Although 42 samples were submitted, the construction of two of the microwave susceptor packages necessitated the analyses of two different portions of the susceptor package. The GC/FID analyses indicated that over 259 individual chromatographic peaks could be separated in the 44 samples tested. All peaks were not found in each sample, nor did they appear in any one sample.

Peak areas were acquired by a Hewlett Packard 3354B Laboratory Data System. The total peak areas for each sample were summed and a total volatiles weight calculated from this sum using the known weight and area counts of the 4-Heptanone internal standard. Justification for quantitating a wide range of compounds while using a FID detector and a single internal standard is supported by published response factors (2). The response factor for 4-Heptanone (ketones) is approximately 0.7. Other classes of compounds with somewhat lower response factors than ketones(0.6-0.8) alcohols, acids(0.2-0.6) and esters(0.4-0.8). These compounds would be slightly underestimated during quantitation. Other classes of compounds would have higher response factor and would be slightly over estimated during quantitation. The quantitation of volatiles based upon 4-Heptanone, should not very by more than a factor of 2, and, in fact, would be much less, for most of the compounds encountered.

Figure 2 summarizes the determination of the total volatiles results. Most of the samples show less than 40 $\mu g/in^2$ total volatiles. Over half of the samples analyzed show total volatiles of less than 10 $\mu g/in^2$. Only seven samples showed total volatile over 40 $\mu g/in^2$.

Analyses by GC/MS. It was also desired that as many of the major volatile peaks as possible be identified. Five microwave susceptor packaging samples were selected for further analyses by GC/MS. These samples were selected for the number of peaks that they had in common with the remaining samples. It was assumed that samples with peaks having similar retention times to those identified by GC/MS would have like identities. This assumption permitted the rapid evaluation of the data.

The samples were analyzed in a manner similar to that described above. The temperature program for the GC was modified to enhance the sensitivity of the GC/MS. To achieve greater sensitivity a greater quantity of volatile material had to be placed on the GC column. This was achieved by cold on-column trapping. Specifically, the capillary chromatographic column was cooled to -20°C. The two ml headspace gas samples were injected into the GC in the splitless mode and the injector was held in this mode for two minutes while the column was maintained at -20°C. After two minutes, the injector was switched to the split mode and the column heated to 30°C at 20°C/min. When the column reached 30°C, the heating program was changed to 4°C/min. This analyses scheme permitted the concentration, resolution, and identification of compounds present with a high degree of confidence.

The resulting GC/MS spectra were compared to the spectral data in the EPA/NIH Mass Spectral Library in a effort to identify the compounds. In addition, reconstructed ion chromatograms were prepared and compared to the original GC/FID chromatograms. Difficulty in identifying some of the peaks is due to three main factors; (1) The peaks contain two or more co-eluting compounds which produce a hybrid mass spectra that cannot be matched with a known compound; (2) The spectrum of the compound of interest is not included in the library; and (3) The mass of material may be insufficient to adequately separate the spectrum from that of the background with any degree of confidence. Also some of the samples when reanalyzed by GC/MS, also showed that a number of peaks previously tabulated as individual substances in different samples were, in fact, the same compound.

Early eluting peaks (peaks 1–20) were difficult to identify due to base line upset caused by the injection of a considerable amount of gases. These peaks have retention times between 2.17 and 4.29 minutes. Comparison of the reconstructed ion chromatograms show that many of the peaks are common to all the samples and that the resolution between peaks is very poor. That, combined

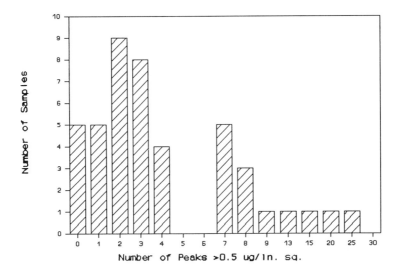

Figure 2. Determination of Total Volatile Results.

with the lack of a sufficient number of mass fragments, has prevented the identification of these peaks by GC/MS. Table VII shows these peak numbers and their suggested identification based on the chromatographic retention time of pure standards.

Table VIII lists those peaks identified by GC/MS. 1,1,1-Trichloroethane was found in seven susceptor samples. This substance in not considered a carcinogen, and in fact is of low toxicity, submitters of these samples have

Table VII. Microwave Volatile Extractives Identified by Retention Time Only

Peak No.	Compound
1	Methane
4	Propyne, Ethane
6	Propylene
7	Propane
9	Isobutane
10	Acetone
11	2,2-Dimethylpropane
14	2-Methylbutane
15	Hydrocarbon
16	Methanol

Table VIII. Microwave Volatile Extractives Identified by GC/MS and/or Retention Time

Peak No.	Compound	No.	Compound
3	Acetaldehyde	101	Furfuraldehyde
18	2-Propenal	102	2-Carboxaldehyde
19	Furan	108	4-Heptanone
20	Ethanol	114	3-Heptanone
21	4-Methyl-8-pentenoic Acid	115	2-Heptanone
26	2-Propanol	117	Heptanal
29	Butanedione	118	2-Butoxyethanol
31	Butanal	124	2-Ethyl-1-hexanal
34	2-Methylpropanol	128	1-(2-Furanyl)-ethanone
35	2-Butanone	130	3,3-Dimethylhexanal
36	2,5-Dihydrofuran	133	2-Heptanal
37	1,1-Dichloroethane	134	2-Ethylhexanal
43	Butanal	141	Decane
44	1,1,1-Trichloroethane	143	Benzaldehyde
48	Formic Acid	144	Undecanone
49	2-Butanal	146	2,2,3,4-Tetramethylpentane
50	2-Butenal	147	Octanal
51	Butanol	149	Dimethyldecene
52	Benzene	161	2-Ethyl-1-hexanol
56	Pentanal	166	Benzeneacetaldehyde
57	2-Propylfuran	171	1-Phenylethanone
58	2-Methyl-2-butanone	174	2,4-Dimethylhexane
59	Acetic Acid	184	Butoxyethoxyethanol
60	3-Methyl-2-butanone	186	Nonanol
64	Pentanal	193	Decanal
66	2-Heptene	195	2-Ethylhexylacetate
67	2-Pentanone	196	Benzoic Acid
74	3-Methylbutanol	200	Aliphatic Hydrocarbon
77	2-Methylfuran	202	3,3-Dimethylheane
78	2-Hydroxypropyl-2-Methyl -2-propionate	206	Decanal
83	Toluene	214	2,4-Dimethyl-1-decene
84	3-Methylheptanone	225	2,4-Dimethyldecane
87	3-Hexanone	231	Aliphatic Hydrocarbon
89	3-Methyleneheptane	243	Dodecanal
91	Hexanal	253	3-Ethyloctane
96	Tetrachloroethylene	254	Aliphatic Hydrocarbon

committed to reduce this substance's presence to the lowest possible level due to the perception that it is associated in some way with other chlorinated solvents about which concern has been expressed.

The presence of benzene was indicated in five samples. One sample was an experimental construction and was never offered commercially.

Two samples were never used commercially. One sample was no longer in production and not presently in the marketplace. The one remaining sample was immediately reformulated to eliminate the presence of benzene.

The data was examined for the presence of 13 specific compounds of toxilogical concern. These compounds were; Methylene Chloride, Trichloroethylene, 1,4-Dioxane, Epichlorohydrin, Carbon Tetrachloride, Ethyl Acrylate, Benzene, Chloroform, 1,2-Dichloroethane, Acrylonitrile, Ethylene Oxide, 2,4-Toluene Diamine, and Vinyl Chloride. With the exception of benzene no other compounds of a toxilogical concern were found.

Conclusions

Suitable methodology was developed and validated for the determination of the amount and nature of substances with the potential of volatilizing from susceptors and migrating into foods under the intended conditions of use. The methodology takes a worst case stance in that it is assumed that 100% of the volatile substances identified migrate to the food. The method sensitivity employed is equivalent to a very small dietary exposure of 0.5-2.5 ppb for detected volatile substances.

Forty two susceptor samples representing a variety of susceptors and food uses were collected and analyzed. Of the identified compounds only one compound, benzene, was found to be of concern. Immediate action by the committee resulted in the elimination of this compound. Eighty two compounds have been identified to date only one of which, benzene, is of concern. Continued efforts by industry of this type will ensure the safety of microwave susceptor packaging.

While the stimulus for this work was primarily concerned with the health and safety of susceptor packaging, the methodology developed will also have an impact on packaged food quality. Flavors contributed to food product by the migration of package components are also being examined by this and similar methodology.

Literature Cited

1. Report to the FDA by The Susceptor Microwave Packaging Committee, "THE DETERMINATION OF VOLATILE SUBSTANCES FROM COMMERCIAL AND DEVELOPMENTAL SUSCEPTOR MICROWAVE FOOD PACKAGING", December 7, 1989.

2. Dietz, W.A., *Journal of Gas Chromatography*, **1967**, vol. 5, pp. 68.

RECEIVED June 28, 1991

Chapter 7

Toward a Threshold of Regulation for Packaging Components

A Feasibility Study

Lester Borodinsky[1]

Division of Food Chemistry and Technology, Center for Food Safety and
Applied Nutrition, Food and Drug Administration, 200 C Street, S.W.,
Washington, DC 20204

The concept of a threshold of regulation (T/R) for components
of food packaging has evolved from a need for a practical
interpretation of the 1958 Food Additives Amendment to the
Federal Food, Drug, and Cosmetic Act, which requires that
materials that may migrate to food from food packaging, even
in very small quantities, be subjected to premarket approval.
The Food and Drug Administration has explored procedures
with which to make consistent decisions on the need to subject
these materials to the full food additive petition review process.
This chapter will describe a pilot study that has used actual
submissions to the Agency to test the feasibility of such a
procedure, including guidelines designed to differentiate potential
T/R situations that would require a food additive petition and
those that would not.

The 1958 Food Additives Amendment to the Federal Food, Drug and
Cosmetic Act defines a food additive as

"...any substance, the intended use of which results or may
reasonably be expected to result, directly or indirectly, in its
becoming a component ... of any food (including any substance
intended for use in producing, manufacturing, packing, processing, preparing, treating, transporting, or holding food..."

Thus, even substances that may migrate to food in very small quantities

---

[1]Current address: Keller and Heckman, Suite 1000, 1150 17th Street, N.W., Washington, DC
20036

from packaging materials have been the subject of regulations. With advances in analytical methods of detection over the years, lower and lower levels of migrating substances are technically subject to food additive regulation. The following question continues to be asked: "Is there some level below which a substance need not be considered the subject of a regulation that specifically permits its use?" The Food and Drug Administration (FDA) has grappled with this "Threshold-of-Regulation" (T/R) question since 1958 for substances that might migrate unintentionally into food. Generally, FDA has handled such situations on a case-by-case basis. Increasingly, however, there have been calls for FDA to develop a T/R policy that would permit the use of components of food-contact articles without a regulation specifically delineating that use.

To develop such a policy, FDA must consider a number of important issues, such as the reliable measurement of low levels of migration, low dose risk assessment, the likely potencies of possible carcinogens, and an array of legal and regulatory constraints. The possible ways of weaving together this array of considerations to arrive at a viable policy for the Threshold of Regulation has been discussed previously (1-4).

In this chapter, the practical matters involved in of the implementation of a T/R policy will be discuss discussed: is such a policy feasible, and are there likely benefits from the implementation of such as policy? To answer these questions, a pilot study was undertaken, and this chapter will focus on this pilot study. FDA recognized early on that if every substance that is evaluated passes, you don't have a viable policy and conversely, if every substance fails, you don't have a policy either.

However, before proceeding to a discussion of the pilot study and its outcome, an overview of the food additive review process will be presented.

Review of Food Additive Petitions

Each component of a food package may be considered to be an indirect food additive, as described in the quote of the 1958 Amendment to the Act which is quoted in the introduction. The Amendment requires the pre-market approval of food additives. This approval, which takes the form of a regulation that permits the use of the food additive and stipulates the conditions under which it may be used, is determined through a multidisciplinary evaluation of a food additive petition.

In Figure 1 is a diagram describing the individual operational units that perform at least one discreet evaluation of each petition. Note that evaluations by the Divisions of Nutrition and Microbiology, which are required on occasion for direct food additives, such as artificial sweeteners, have been omitted from this diagram because they are rarely needed for indirect food additives. In addition, although included in this diagram, evaluations by the Cancer Assessment Committee (CAC) and the Quantitative Risk Assessment Committee (QRAC) are not needed in all cases, only those involving carcinogenic impurities. A number of different operational units evaluate each petition. The evaluation typically involves more than one individual. For example, the technical evaluation memoranda written by the chemist (in the Division of Food Chemistry and Technology [DFCT]), the toxicologist (in the

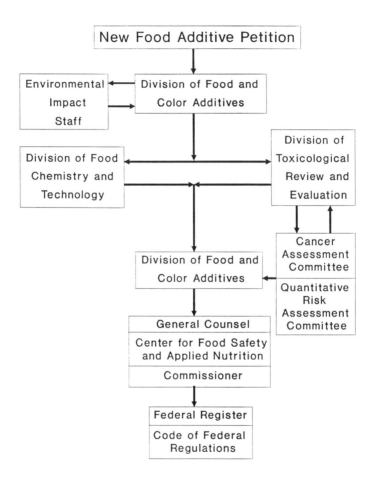

Figure 1.   Operational units at the Food and Drug Administration that evaluate indirect food additive petitions.

Division of Toxicological Review and Evaluation [DTRE]) and the environmental reviewer (of the Environmental Impact Staff [EIS]) must each be
deemed acceptable by the reviewer's immediate supervisor, and in some cases,
by one or more supervisors above that level.   These memoranda are
addressed to the attention of a Consumer Safety Officer (in the Division of
Food and Color Additives [DFCA]), who managers the petition and has
his/her own set of supervisors.
      Although some of these reviews occur concurrently using separate copies
of the petition, the petition passes through the hands of many individuals; a
considerable amount of time may pass before the process is concluded and a
regulation appears in print. It has been estimated that for a "perfect" petition
(i.e., one which has not raised any questions from any of the reviewers
regarding the adequacy of the data submitted in the petition) it will still take
between 6 months and a year to complete the review and arrive at a
regulation. Of course, if deficiencies in the petition are identified, the time
until a regulation is issued can be lengthened considerably, 1½-2 years, or
longer. This time span reflects FDA's time to review the petition, and also
the time spent by the petitioner to respond to any questions raised during the
review.

T/R Consideration

Before discussing the pilot study, some of the salient points impinging on T/R
considerations will be summarized.

> One approach that has been proposed is to base the T/R policy on
> potential carcinogenic risk. Known or suspected carcinogens will be
> excluded from this policy.

> The potencies of known or suspected carcinogens appear to distribute
> in a Gaussian fashion (Figure 2).

> Using this potency distribution and an assumption of the linear
> proportionality of carcinogenic risk as a function of dose, a Gaussian
> distribution of a "10^{-6}-risk-equivalent" is created (Figure 3).

> This distribution reveals, for example, a probability of approximately
> 50% that the human intake of a carcinogen at a dietary level of 1 ppb
> could present less than an upper-bound lifetime risk of cancer of one-
> in-a-million. At a dietary level of 0.2 ppb, the probability of less than
> a one-in-a-million lifetime risk is approximately 70%.

> In order to estimate probable dietary levels of the substance in question
> and, thus, relate intake levels to the "one-in-a-million" risk-equivalent,
> data are needed for evaluation. These data include food-contact use
> (time, temperature, food-type, repeat-use vs single-service-use, use level,
> etc.) and data to estimate migration to food. The latter may be either
> validated migration data with food or food simulants, or data on which
> to base 100% migration to food.

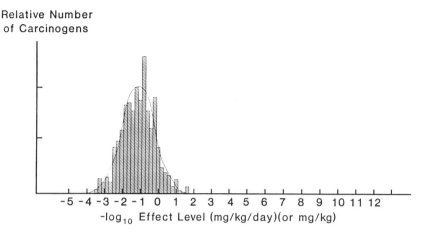

Figure 2. Probability distribution of potencies of known or suspected carcinogens. (Adapted from reference *2*).

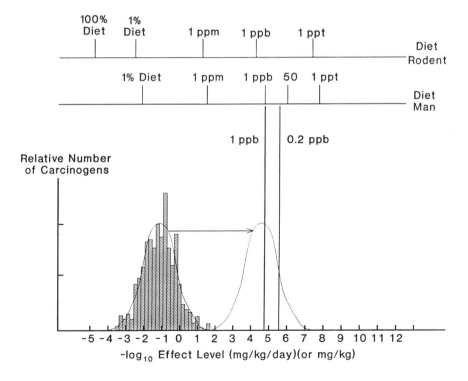

Figure 3. Probability distribution of potencies of known or suspected carcinogens shifted to a distribution of a "10^{-6}-risk-equivalent" by assuming a linear proportionality of carcinogenic risk as a function of dose. (Adapted from reference *2*).

T/R Operational Guidelines

Considering the above points as the basis for implementation of a T/R policy, a committee of individuals with considerable experience in evaluating food additive petitions developed guidelines to be used by a three-person Threshold-of-Regulation Panel (TRP) to test the feasibility of a T/R policy. The guidelines were organized into a three-step process. In the <u>Preliminary Screening</u> step, food-contact situations deemed "unsuitable" for this type of review would be identified, removed from the loop and evaluated in the usual way. Each situation not rejected in the first step would proceed to the <u>Evaluation</u> step, in which the proposed food-contact situation would be judged as to whether it qualifies for exemption from regulation because it is below the T/R. In the <u>Disposition and Record-keeping</u> step, a permanent record of the findings of the TRP would be created and the requestor informed of the Center's decision.

Each of these steps is described in greater detail in the following sections.

Preliminary Screening. During this screening step, several questions were posed, including:

- Is the food-contact situation well-defined enough in terms of use conditions? If not, then the TRP must decide whether or not correspondence with the firm will be undertaken.

- Is there sufficient information to permit a calculation of likely human exposure or worst-case exposure? The latter is usually based on 100% migration? If information is insufficient, then the TRP must again decide whether or not correspondence with the firm will be undertaken.

- Is the subject material a known or suspect carcinogen? Is the subject material likely to contain as an impurity a known or suspect carcinogen? If so, then the situation fails the screening step.

- Is it likely that this review will require substantial correspondence between the requestor and FDA? If so, then the situation fails the screening step.

Evaluation. The requests not rejected based on the foregoing questions underwent actual evaluation; in many respects, this evaluation was similar to that performed during the review of a food additive petition. The T/R evaluation involved the following:

- The dietary intake of the substance was estimated from migration data or other information included in the submission, or from other information readily available.

- If toxicological data were included with the submission, or if such data existed in FDA files, it was determined whether the data support the estimated dietary intake.

- FDA's precedent files were examined to determine whether existing precedents might influence a judgment on the subject situation.

- Finally, the estimated dietary intake was compared to the nominal T/R dietary "gray zone" of 0.1-1 ppb. Because the Agency has not yet established an actual threshold by rulemaking, the estimated dietary intake was placed in one of the following categories: those that clearly failed (i.e., above 1 ppb), those that clearly passed (i.e., below 0.1 ppb), and those that would also pass if the threshold is established as high as 1 ppb.

Final Disposition and Record-keeping. Once a decision on passing and failing is rendered, the following activities occur:

- Correspondence describing the TRP's decision is drafted for signature by the appropriate official (Director of the Division of Food and Color Additives, Director of the Office of Compliance, etc.).

- The submission and resulting correspondence are recorded in a reference database to establish a T/R precedence file.

- All time (person-hours) expended in the effort is recorded.

Testing the Guidelines

The three-person Threshold of Regulation Panel consisted of one toxicologist (from the Division of Toxicological Review and Evaluation), one chemist (from the Division of Food Chemistry and Technology) and one Consumer Safety Officer (from the Division of Food and Color Additives).

The TRP met approximately once every 2 weeks over a period of 5 months. The 9 convened deliberation sessions lasted a total of 8 hours. In these sessions, 25 "real" situations were evaluated; because some of these requests involved more than one substance, a total of 35 decisions were rendered.

Of these 35 decisions, 12 failed, 13 passed and 10 received "gray-zone" (0.1-1 ppb) passes (Figure 4). Of the 12 that failed, 4 needed more information, 2 would have required too much time to fully evaluate, 1 was a suspected carcinogen and 5 were rejected on the basis of dietary estimates. Of the 13 that passed, 11 were based on qualitative evaluations (e.g., repeat-use situation, or the known nature of the subject material) and 2 were based quantitatively on dietary estimates. The 10 that received "gray-zone" passes had dietary estimates ranging from 0.3 ppb to 1 ppb. In each of the 3 decision categories (i.e., failed, "fully" passed, "gray zone" passed), estimates were based on actual extraction data as well as on worst-case (100%) migration (based on use level).

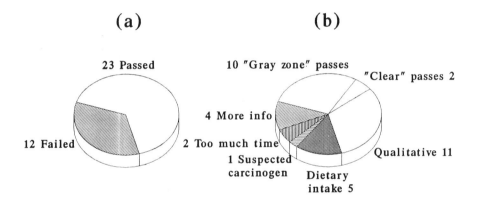

Figure 4. Results of the 35 decisions rendered during the course of the pilot study. (a) Overall pass-fail distribution. (b) Detailed pass-fail distribution.

In addition to the 8 hours in which the TRP was convened (corresponding to 24 person-hours), the additional review time expended was as follows:

Division of Toxicological Review and Evaluation	6.4 hours
Division of Food Chemistry and Technology	63.0 hours
Division of Food and Color Additives	<u>26.3 hours</u>
	95.7 hours

The total time deliberated was, therefore, 119.7 hours for 35 decisions, or 3.4 hours/decision.

Conclusions

A strict record was not kept on the turn-around time for each decision, i.e., the time interval between the receipt by the TRP of the submission and the rendering of the decision. Although records of this type were not kept, we estimate that the average turn-around time was probably approximately 3 months, including 2-3 weeks to prepare and issue a letter to the requestor. This may be compared to an interval of 1-2 years for the approval of an indirect additive requested in a food additive petition. This shorter turn-around time would clearly be a benefit to those in the industry whose materials, under well-defined conditions of use, could be evaluated with the implementation of a T/R policy.

The results of the study demonstrate that the guidelines can be used to discriminate between those situations that could pass a T/R and those requiring the submission of a food additive petition.

Finally, and most importantly, the average evaluation time is only 3.4 person-hours/decision, which is considerably lower than the average number of person-hours expended by the Agency (including legal review of draft documents and Federal Register publication) for each food additive petition. This could be of considerable benefit to the FDA and to actual petitioners in terms of allocation of available resources.

Literature Cited

1 Flamm, W.G.; Lake, L.R.; Lorentzen, R.J.; Rulis, A.M.; Schwartz, P.S.; Troxell, T.C. In *Contemporary Issues in Risk Assessment;* Whipple, C., Ed.; Plenum Press, New York, 1987, pp 87-92.

2 Rulis, A.M. In *Risk Assessment in Setting National Priorities;* Bonin, J.J.; Stevenson, D.E., Eds.; Plenum Publishing Corporation, New York, 1989, pp 271-278.

3 Rulis, A.M. In *Food Protection Technology;* Felix, C.W., Ed.; Lewis Publishers, Inc., New York, 1987, pp 29-37.

4 Munro, I.C. *Regulatory Toxicology and Pharmacology* **1990,** *12,* 2-12.

RECEIVED June 20, 1991

Chapter 8

Packaging Industries and the Food-Additives Amendment

A Reprise

Jerome H. Heckman

Keller and Heckman, Suite 1000, 1150 17th Street, N.W., Washington, DC 20036

On September 14, 1966, the author gave a paper entitled "The Packaging Industries and the Food Additives Amendment of 1958 — It's Time for a Change in the Law" at an ACS Symposium on Safety Evaluation of Coatings and Plastics for Food Packaging held in New York City. There have been no significant legislative changes since then, but the ever increasing capability for finding molecular amounts of substances by the application of sophisticated analytical techniques has increased the burdens the 1958 law imposes on those who must comply. In this paper, the 1966 discussion will be reprised, the evolution of some current problems will be discussed (e.g. those created by state legislation, and those that have arisen from public "chemophobia"), and the author will again urge legislative and administrative reforms.

The last time I spoke at an American Chemical Society event of this type was in 1966. This is when Jack Frawley of Hercules Corporation and I, among others, delivered papers at a Symposium on Safety Evaluation of Coatings and Plastics for Food Packaging held in New York City at the Commodore Hotel. Our papers were published in the Symposium proceedings. Mine, entitled "The Food Additive Amendments of 1958 — It's Time for a Change in the Law," was published in the Food Drug and Cosmetic Law Journal later that year.

Jack's paper proved seminal in scientific circles. It ultimately led to an important National Academy of Sciences study and publication entitled "Quantitative Guidelines for Toxicologically Insignificant Levels of Chemical Additives." Unhappily, neither paper has yet had the desired effect of impacting permanently on the way the Food and Drug Administration (FDA) carries out its responsibilities, or in bringing about regulatory or statutory changes. It has since occurred to me many times that perhaps our papers have done little real good because we were only beginning to focus on the issue of what is significant from a public health standpoint. We had begun our Odyssey into the "never-never land" of trying to define "zero" in the public health sense, but much more needed to happen before the concept of a regulatory threshold could be sold to a nervous bureaucracy.

In the 24 years since that 1966 Symposium, the Commodore Hotel has disappeared, there is probably no one here who was there, and the concepts we advanced may well have been forgotten by all but us. I have spent those years studying the evolution of my old nemesis "public health zero," and have concluded, after a fast rereading, that what we said in 1966 remains good sense today. I have also noted with interest, however, that at least three separate disciplines — chemistry, toxicology and law — must yet come together in government circles, not just industry circles, to make something useful happen. It is, it seems to me, time that we "got on with it." For this reason, I decided to spend my time before you in reviewing some of the bidding and updating it.

A look at the history of this subject will be helpful to put things in perspective. When the ACS Symposium was held in 1966, a number of tremors in the world of packaging regulation had already occurred that complicated a basically very simple regulatory issue. These tremors were causing considerable pain and continue to do so today. My thesis at the time was and is, today, that the use of excessive amounts of FDA and industry money and "sweat equity" to deal in such detail with innocuous packaging component questions is Lilliputian, to be generous. How did it happen?

Prior to 1958, packaging materials issues had commanded some FDA and United States Department of Agriculture attention but the handling of the subject was informal. Strangely enough, the primary handler of the questions raised was a renowned toxicologist, Dr. Arnold Lehman of FDA. His decision making left little to be desired. His actions were prompt, fair, and generally consistent with his and others' (including my own) perception of the very limited potential of an effective packaging material to give rise to a public health problem.

What occurred then, as it does now, is that food processors generally insisted that their suppliers provide assurance that their materials would not adulterate food as a result of unwanted migration, or migration in any amount that would raise Agency anxiety because of public health considerations. To obtain the best possible basis for providing such assurance to their customers, packaging materials suppliers voluntarily corresponded with Dr. Lehman and Robert M. Philbeck of the Department of Agriculture. They submitted their confidential formulas, such extraction data as might be needed, and such toxicology data as were available. In all but the rare case where some serious toxicological concern prompted Dr. Lehman to ask for more study, a letter of "no objection" in the case of FDA or one of "chemical acceptability" in USDA's case, came back to the packaging materials supplier in two or three weeks or less.

As a scientist whose discipline was toxicology, Dr. Lehman probably called upon his chemistry advisers like Dr. Lessel Ramsey for help at times. In any case, it was quite clear to everyone that Dr. Lehman did not demand data in a vacuum and had a good appreciation for the significance of extraction studies in his evaluations. In short, he was a toxicologist of the old school who fully appreciated the fact that the dose makes the poison — thus, he established his own thresholds of regulation, his own sound concept of what might be toxicologically significant. (As he would probably say if he were here today, this was before the damned lawyers got into the act and befouled the situation by imposing a new Statute and inhibiting pure scientific evaluation and common-sense responses.)

To apply such a threshold, it was essential that the Agency be given the extraction information Dr. Lehman relied upon. The FDA scientists, therefore, devised a more or less standard test to be used in collecting the needed data. It came to be called the "spit-in-the-ocean" test. Typically, what one did was to immerse a film sample of one milligram in a one-liter container of solvent and stir the system periodically while the film remained in the solution at 212° for a week. After the prescribed time, the sample was removed and examined. If there was some amount of the film intact, you could conclude that the solubility was less than one part per million and, hence, that there was no potential for migration worthy of public health concern.

This methodology was primitive by today's standards. It certainly would not provide all of the answers for today's packaging superstars like retortable pouches and susceptor microwave containers. However, it was elegant enough for most of the purposes of the times. Moreover, the Lehman and Philbeck letters of that era remain important because any substance deemed "not objectionable" by them later achieved legal immortality under the Food Additives Amendment of 1958.

A showing of one of their letters is about the only basis upon which the present FDA staff will agree that a substance is prior sanctioned within the meaning of Section 201(s) of the Act, and, therefore, exempt from the need for filing a food additive petition under the provisions of the Food Additives Amendment. A prior sanctioned substance is one whose use is sanctioned by an FDA or USDA letter or memorandum issued prior to the enactment of the Food Additives Amendment in 1958. Of perhaps even more significance is the simple fact that, to the best of my knowledge, no substance characterized as safe or unobjectionable by Dr. Lehman or Mr. Philbeck was ever given rise to a public health concern of any type. Why then did someone decide to fix what was not broken?

In 1957 and 1958, some five years after a New York Congressman named James J. Delaney began his hearings into the evils of food additives, the House of Representatives Committee on Interstate and Foreign Commerce conducted the hearing process which was to lead to the passage of the 1958 Food Additives Amendment. The concerns of the packaging industries were expressed to the Committee by The Society of the Plastics industry, the American Pulp and Paper Association, the Adhesives Manufacturers' Association, and the Waxed Paper Institute, Inc. However, the Congress was convinced that the packaging issue was of little import. The focal point of their attention was the question of how to regulate intentional direct food additives. The leading player for the industry side was the Manufacturing Chemists Association (now the Chemical Manufacturers Association).

Actually, even some of the packaging industries most informed leaders were lulled into believing the new law would bring no serious problems — they believed and were assured by FDA spokesmen in the presence of inquiring Congressman (but, unfortunately, not on the record) that the definitional language "may reasonably be expected to become a component of food" would make it possible for the simple, straightforward "ask Dr. Lehman for a 'no objection' letter" approach to remain viable indefinitely. On the other hand, the record does show that there were those of us who protested vigorously against including packaging materials in a complex and time consuming regulatory scheme designed to require the sort of mountains of data one must amass to clear an Aspartame or an Olestra.

In 1958, the new law was passed. Our efforts to have packaging taken out of the food additive definition and regulated separately in some more appropriate way had proven fruitless except in two respects. First, the phrase "reasonably be expected" was inserted so that the law would not prescribe food additive treatment for everything that might somehow come into contact with food; this put a very loose boundary on what FDA could call a food additive. Second, a legislative record was made to support the law's inapplicability to ordinary housewares. It should be clearly understood that the fight to get packaging materials out of the food additives bill was not lost because the legislators paid attention to the problem and consciously rejected our arguments. The struggle to get some agreement on public health zero began with the enactment of this law but it was not because Congress wanted the struggle to take place.

The fact is that the legislators were preoccupied with the perceived need to do something about the uncleared direct additives that were being put into the food supply quite deliberately. They and the chemical industry were also interested in avoiding a specialized ban on anything that might even sound like a carcinogen since Congressman Delaney was insisting there be such a ban. The FDA position was that there was no need for such a special provision since the power and duty to bar carcinogens in the food supply is implicit in the general safety clause of the Food, Drug and Cosmetic Act of 1938, as amended.

In the Spring of 1958, representatives of the Manufacturing Chemists Association told us they were ready to agree to the slightly revised FDA version of a bill. They said they would no longer stand in opposition because of their fear that failure to support the FDA bill at once would lead to enactment of a Delaney Clause. After their support was signalled — and we were told that our opposition was considered of no real consequence since the impact of the bill on packaging was expected to be minimal — the greatest irony of all was that James Delaney had his way by having the Bill amended to include the clause which bears his name on the Floor of the House while it was being debated just prior to passage.

What the direct additives interests most feared had come to pass anyway. At the same time, what had also come to pass — despite the legislators' good intentions and the FDAers' misleading promises of reasonableness — was a law that would unreasonably impact packaging. We lost without a real understanding of what had been done — this remains a sore subject to this day. Indeed, those who have to deal with FDA on packaging matters are continuing to pay a high price for the Congressional inattention to detail.

I will not burden you once more with an oral litany of how the FDA for two years kept its promises to Congress and acted reasonably on the "expected to become a component of food" or threshold of regulation issue and then departed on a frolic of its own for the next 30 years. For those who want to know more of the story of the years through 1966, my earlier ACS paper is published in the December, 1966, issue of Food Drug Cosmetic Law Journal beginning on page 647. For those who might like their memories refreshed on the events that have occurred since 1966, I have prepared as an attachment for this paper a chronology of the highlights of this entire saga as it has played out between 1960 and today. Suffice it to say that FDA's departure from rationality in 1960 has led all of us interested in packaging regulation to engage in a tenacious effort to find the Grail of a threshold of regulation ever since.

This search has centered on a seemingly simple idea — that there must be some level of exposure to substances, or at least most substances, that presents no real threat to public health and, hence, can be used to save the effort that currently goes into the full scale Food Additive Petition treatment of materials. This treatment now requires an average of one to two years to bring forth a dispositive Agency action.

We have all been looking for this public health zero ever since at least 1960. Indeed, as readers of the attached chronology will duly note, in

1968, FDA's then Associate Director of Regulatory Programs in the Office of Compliance proposed that such a level be set for everything but known carcinogens, heavy metals, or substances proven to be acutely toxic at levels below 40 parts per million (ppm). This proposal, called the "Ramsey Proposal" after its official author, looked towards eliminating the need for the filing of Food Additive Petitions for adhesives, components of articles made for repeated use, components of paper packaging for dry foods, and any substance which is not a carcinogen, a heavy metal, or an acutely toxic substance if it will not migrate to food in an amount to exceed 50 parts per billion (ppb). No scientific reason has ever been given for not moving forward with the Ramsey Proposal as an excellent start on establishing one threshold of regulation. Why then has so much time passed without any constructive action having been taken?

One of the primary reasons we have not solved the problem, and moved forward in a way that is conceded by all to be needed, is that everyone keeps insisting that the fault lies with those in every discipline other than his own. The "buck passing" is classic. The chemists among us say it is not their problem — their job is to develop ever more sensitive methodology so that we can now measure some substances' presence in parts per quadrillion. That this makes it more difficult to say that nothing will become a component of food in amounts of any significance is not their concern. Moreover, many of them will tell you that the lawyers, toxicologists, or some other policymaking discipline should decide where "zero" is. This, they contend, is not a chemist's job, even if he or she is also among the FDA policymakers who do decide what does and does not require a Food Additive Petition.

The toxicologists have sidestepped the "zero" question by generally taking the position that they can only say something is safe when they have reams of data on the substance. They maintain this view even if the material is used at such low levels that the sort of the data they might like will never be available unless the government — you and I and the rest of the citizenry — pays the tab for it. The idea of classifying according to exposure data and structural information has been accepted grudgingly, if at all, until relatively recently. This is the case even though the information always had to be on hand since no studies would ever have been done on any food additive without advice from the Agency's scientists as to what they would require. This is how the world-famous Redbook came to be and now sets forth the kind of data one must present in a Food Additive Petition, depending entirely on the chemist's estimates of dietary exposure and the nature of the chemical to be studied.

The lawyers, including some of FDA's distinguished past Chief Counsels, have generally taken the position that the problem is not really important since the law permits businessmen to make their own judgments on where the public health zero is. They will tell you that anyone has the right to decide that his product may not reasonably be expected to become a component of food and, hence, need not be subjected to FDA's extremely cumbersome and clumsy petition procedures. (They will also tell you that pinning down "zero" for regulatory purposes is primarily a scientific issue, not a legal one.) The only problem with this position is that most of the lawyers do not have to sell the product to a customer who insists that he will not buy unless he can see in some black and white document issued by the government that the product he is being asked to buy and use is cleared by FDA or is exempt from the need for clearance.

In my view, it serves no worthwhile purpose to continue the buck passing. Beginning in 1985, calls were made by FDA's Center for Food Safety and Nutrition, and then-Commissioner Frank Young for the establishment of a threshold of regulation, at least for packaging materials. Indeed, since 1977, FDA has had before it a Citizen's Petition (7CP3313) urging the Agency to adopt a reasonable threshold policy for indirect additives. With the work already done in this area by FDA's Dr. Alan Rulis and the Canadian Centre for Toxicology (CCT), as reported on in the study it did for The Society of the Plastics Industry, it is time for action. The threshold project called for CCT to convene a panel of renowned independent experts to examine the scientific concepts for evaluation of trivial, or de minimis, carcinogenic and non-carcinogenic risks resulting from low level exposure to food-contact substances. The overall conclusion of the Panel is that, for indirect additives for which no toxicological data have been developed, a threshold of one part per billion (ppb) in the diet is appropriate. For non-genotoxic substances, the Panel recommends a regulatory threshold in the range of 5-10 ppb in the diet. For most packaging materials this threshold translates to a 50-100 ppb migration into appropriate food-simulating solvents. There is no longer any reason not to have the 1968 Ramsey Proposal put before the public for full-scale comment and ultimate adoption. This is the way to find out if there really is any continuing dissent. If not, then we can put an end to the search for what we already had in 1958, a rational threshold of regulation that made for simple treatment of the simple problem that passing on food contact materials safety really is.

As the chronology attached to this paper indicates, it would appear that we cannot at this time bring about a change in the law because it is too difficult to get legislative attention for an industry problem that does not have much vote-getting glamor on election day. But if we are all agreed that wasting of government and industry resources is, now more than ever, to be avoided and FDA believes that its leaders have been saying about the need for a threshold of regulation, we can go forward.

The stage is set for action if FDA wants to take it. The very existence of the Redbook shows that there are valid general principles in toxicology; the Agency chemists should not resist the drawing of sensible lines since this will not prejudice their right to develop even more refined detection methodology if anyone needs it; and though some lawyers might privately object to the providing of more certain regulatory criteria and principles on economic grounds, I can assure you that none will publicly oppose the action I advocate here.

All that is needed is a well measured response to the pending Citizen's Petition 7CP3313 filed in 1977. And all FDA has to do to establish what is now a scientifically justified public health zero for food contact materials is to adopt the principle Dr. Ramsey put forward for industry consideration in 1968, two years after the first major call for action took place at another American Chemical Society Symposium like this one.

APPENDIX

The Saga of the "No Migration" Controversy

1960-1961. Prior to this time, under policies established immediately after the Food Additives Amendment of 1958 became effective, the FDA invited manufacturers of substances used in packaging materials or processing equipment applications to write to the Agency, supply data relating to the question of whether or not the substance might be expected to migrate to food, and solicit the Agency's opinion on whether or not the substance should be dealt with as a food additive problem. FDA indicated that such letters would be dealt with promptly by the making of a suitable decision and the issuing of letters providing the FDA opinion as to whether or not the substance might be considered a "non-additive" on the grounds that it was not reasonably expected to become a component of food. Indeed, the Agency's food additive regulations required that such advice be given (21 C.F.R. Section 121.2 and 121.3(d)). Such letters were issued routinely, and were usually signed by the designated chief administrator of the food additives regulatory program, Arthur A. Checci.

This was the regular procedure followed until the time of the annual FDA-Food and Drug Law Institute meeting which took place in the Fall of 1960. At that time — and for the first time in our recollection — then Assistant Commissioner J. Kenneth Kirk announced that the Agency would no longer issue such letters. The rationale provided for this decision was simply that the Agency considered it an undue burden to respond to the letters and was also fearful that providing them might give some suppliers a marketplace advantage over others who might not be able to afford to carry out suitable extraction studies. (At the time, such a study probably cost no more than $2,000 or $3,000).

When pressed for a legal rationale for this decision, Assistant Commissioner Kirk called upon then Deputy Commissioner John L. Harvey, himself a lawyer, who provided his legal opinion that:

> We . . . reevaluated our position . . . and came to the conclusion that basically, if there was enough reason to run extraction studies on packaging or equipment materials, why shouldn't it be concluded that it would be reasonable to expect that the substances involved would, in fact, become a part of the food? Since the law refers to "reasonably to be expected," we then began to advise those who asked that we were not in a position to give them a letter which would absolve their product from any responsibility from under the Food Additives Amendment but instead suggested that they file petitions. That is the present status of this item. *(1)*

On this premise, frequently characterized as "right out of Alice in Wonderland," the FDA took the position that the only way it would rule on the acceptability of the substance would be if a full-scale Food Additive Petition were filed leading to a regulatory clearance. Since Mr. Kirk reiterated this position as the position of the FDA again at the next FDA-Food and Drug Law Institute session in 1961, these dates are appropriate to use as the marker dates for the beginning of a now 30-year-old controversy over the jurisdictional question.

The policy adopted in 1960 and 1961 has led to the filing of innumerable Food Additive Petitions in situations where both the Agency and industry could have been spared the time and expense necessary to handle such an effort if a simple jurisdictional decision were made at the threshold.

September 14, 1966: This is the date of the American Chemical Society Symposium on Safety Evaluation of Coatings and Plastics for Food Packaging held in New York City. The papers presented at that session were extremely critical of the FDA and especially its policy with respect to the so-called "no migration" cases. One of the two most critical papers was presented by Dr. John P. Frawley of Hercules who set forth an interesting proposed remedy for the situation by suggesting that any substance used at less than 0.2% in an indirect additive application should be exempt from the need for petitioning unless the substance was a known carcinogen, a pesticide characterized as an economic poison, or a substance proven to be toxic at a level of 40 ppm or less. At this same symposium, Jerome H. Heckman presented a paper entitled "The Packaging Industries and the Food Additives Amendment of 1958 — It's Time for a Change in the Law." The thrust of this paper was that the FDA's entire system for dealing with indirect additives, and particularly those that might not reasonably be expected to become a component of food, was a travesty that needed to be remedied by Congressional action if the FDA would not act on the matter. Both of these papers are still referred to quite frequently.

February 13, 1967: Movements were underway to seek the appointment of a Government Industry Advisory Committee under Executive Order 11007 to consider the entire matter of the Agency's procedures with respect to indirect Food Additive Petitions. At about that same time, in a letter to The Society of the Plastics Industry, Inc. (SPI) dated February 9, 1967, then Commissioner James L. Goddard indicated that he wanted to " . . . assure you that we are most anxious to do everything possible to simplify this procedure . . ." and he advised that he was certain that the entire situation would be improved. No Government Industry Advisory Committee was ever appointed nor was there any follow-through on the Goddard promise.

August, 1967. At a hearing of the House Select Small Business Subcommittee on Regulatory Agencies, chaired by the Hon. John D. Dingell of Michigan, Dr. Goddard was asked a series of questions about the FDA's regulation of indirect additives and its interpretation of the part of the statute relating to the question of when a substance is an indirect additive and when it is a "non-additive." Since the questions were asked in the context of a hearing that dealt with a variety of other matters, Dr. Goddard requested 30 days in which to respond in writing to the questions raised by Mr. Dingell.

No such response was ever sent to the Subcommittee. Instead, the FDA called a "National Conference on Indirect Additives" and used the calling of the conference as a means of assuring Congressman Dingell that it would report back to him later on his questions, presumably after the Conference was held and data was collected that might be of value in this respect.

At the same time, the Agency was considering changes in its procedural regulations and extensive comments were being received on this proposal. Those interested in indirect additives thought that the procedural regulation comment opportunity might be used to induce FDA to improve the procedures it used in regulating indirect food additives.

During 1967, a special committee of the National Academy of Sciences reviewed the Frawley paper given at the American Chemical Society Symposium and ultimately produced a document entitled "Quantitative Guidelines for Toxicologically Insignificant Levels of Chemical Additives." This document was published and recommended strongly that the FDA and other agencies recognize the fact that any substance which constituted less than 0.1 ppm in the diet, other than a known carcinogen, a pesticide or a heavy metal, presents no public health problem.

February 13-14, 1968. On February 13-14, 1968, at the call of Commissioner James Goddard who opened the meetings, the "National Conference on Indirect Food Additives" was held to discuss all issues bearing on the perceived procedural and other deficiencies in FDA's administration of the law. The two day conference, held at the State Department and extremely well attended, included presentations by Mr. George W. Ingle, then chairman of SPI's Food, Drug, Cosmetic Packaging Materials Committee who expressed the Society's concerns regarding FDA's requirements that food additive petitions had to be filed for non-migrants, ergo, "non-additives," and Dr. Ramsey, who addressed the "Scientific Aspects of the Migration Problem."

1968. Dr. Ramsey, then Associate Director for Regulatory Programs in the Office of Compliance of the FDA, presented a paper indicating that the FDA had under serious consideration a proposal which would exempt from the requirement for the filing of a Food Additive Petition all substances used in minor amounts in adhesives, paper packaging for dry food, and all repeated-use applications.   At the same time, he indicated that the Agency was considering amending then Section 121.2500 of the Food Additive Regulations to also exempt any substance the migration of which into the usual food-simulating solvents under the prescribed FDA test conditions would not exceed 50 ppb. The exemptions Dr. Ramsey proposed would be for all substance used in the repeated-use, adhesives, dry food and less than 50 ppb situations unless the substances were heavy metals, known carcinogens or had been shown to be toxic at 40 ppm or less. Some characterized this Ramsey Proposal, as it came to be known, as a loose embodiment of the Frawley concept of 1966. However, it proposed a much more stringent parts per billion pass-fail test than the Frawley paper.

May 6, 1969. The Ramsey Proposal was sent by FDA to all of the major associations that participated in the National Conference on Indirect Food Additives held in 1968 with a request for comments. The associations included the National Flexible Packaging Association, American Paper Institute, American Petroleum Institute, the Soap and Detergents Manufacturers Association, the Can Manufacturers Institute, the Adhesive Manufacturers Association, the Aluminum Association and The Society of the Plastics Industry, Inc.

June 3, 1969. At a meeting held at about this time, an ad hoc committee of representatives of the above-named associations was appointed to negotiate with the FDA on the Ramsey Proposal. During this period, it was generally believed that the 50 ppb criterion was virtually useless because the state of analytical chemistry at the time purportedly did not permit testing at such a low level, or so it seemed to those directly involved in the subject. The Committee was generally charged with the task of attempting to persuade the FDA that a 0.5 part per million level should be an adequate way in which to draw the line since this would permit some materials to pass under the testing procedures known to be available then. Also, this was a level that was considerably below that recommended by Dr. Frawley and, therefore, appeared to be quite conservative. Unfortunately, this argument was unacceptable to the FDA — although why this was so was never made completely clear — nor was any definitive action taken in response to the comments made as Dr. Ramsey had requested.

February 26, 1970. At a meeting on this date, Dr. Ramsey advised a representative of the packaging group that it was unlikely that action would be taken on his proposal for some time because the FDA was preoccupied with its concerns about cyclamates then, and was also "settling in" under a new Commissioner (Edwards) who had just been appointed.

August 21, 1970. At this time, Mr. Thomas C. Brown, then Director of the Office of Compliance of the Bureau of Foods of the FDA, when asked in a letter by Jerome H. Heckman as to the Agency's policy on issuing opinions in "no migration" cases, indicated that he did not concur in the view taken by Assistant Commissioner Kirk in 1960-61 and that, therefore, the Agency would again issue letters in response to appropriate requests with respect to the matter of whether or not a substance is a food additive within the definition in the Food Additives Amendment. Such responses would be based on Agency review of suitable migration data. This widely-hailed "Tom Brown letter" gave industry some hope that the pre-1960 fair interpretation policy would be reinstituted.

It is believed that some letters did issue and a few probably continue to be issued in the wake of the Tom Brown letter, but FDA has in recent years given such letters the pejorative nickname of "giveaway letters." Hence, they are rarely given those who request a dispositive opinion; instead, companies are always advised "to file a Food Additive Petition." The clear preference of the Agency staff is to advise all inquirers that "they should file a Petition" with respect to any request for a status opinion.

January 13, 1971. At a public meeting, Director of the Office of Compliance Tom Brown indicated that Dr. Ramsey was working on a proposal to deal with some of the industry complaints and that such a proposal "would be out soon in a formal way."

June 3, 1971. At this time, in another session with Dr. Ramsey, representatives of the packaging group were informed that the Agency could see no way to go forward with the Ramsey Proposal even though it continued to believe it was scientifically sound. It was not made entirely clear as to why the Agency believed it could not move forward, although there were rumblings about "political difficulties," and lack of understanding on the part of some of the Agency toxicologists.

April 1, 1977. During all of the time from 1970 to 1977 the Bureau of Foods of FDA resisted giving "no migration" letters as much as possible, although some were issued when the case was pressed vigorously at higher levels. For example, on June 5, 1972, Richard J. Ronk of the FDA provided a letter clearing a methacrylonitrile styrene resin for a soft drink bottle on the basis of data indicting that there would be no migration of any uncleared component of the bottle to food when the bottle was tested with a method sensitive to 50 ppb. This letter was subsequently supplanted by a Food

Additive Petition on a similar substance (acrylonitrile styrene) and, therefore, was only remarkable for the fact that such a letter was issued in a case involving a very visible application, but only after an appeal was made to the Agency's Chief Counsel.

By 1977, the FDA had certainly returned to its older policy of refusing to issue letters and urging everyone to file a Food Additive Petition in almost every sort of case. As a result of this, and a public release issued by the FDA in a celebrated soft drink bottle case in 1977, at the suggestion of the then General Counsel, Richard A. Merrill, a Citizens Petition was filed by the Society of the Plastics Industry, Inc. with the FDA on April 1, 1977. For all practical purposes, the Petition simply solicited formal adoption of a more demanding version of the Ramsey Proposal of 1969 and urged that it be adopted as a way of clarifying a very confused situation. This Petition was assigned the Citizens Petition No. 7CP3313 by FDA and was docketed at 77P-0122. No further word has ever been received regarding this Petition, no disposition of it has ever been made, and it continues to remain pending without action at the FDA, despite the fact that the filing of it was suggested by the Agency's Chief Counsel.

November, 1979. The United States Court of Appeals for the District of Columbia Circuit rejects FDA's expansive jurisdictional assertion that a substance meets the Act's food additive definition merely because it is present in a food contact material. Monsanto v. Kennedy, 613 F.2d 947 (D.C. Cir. 1979). The Court ruled that, before a substance meets the component element of Section 201(s) of the Act (a food additive is one "reasonably expected to become a component of food"), FDA cannot simply rely on the fact that the substance is present in the package, but must have evidence that its presence in food can be predicted on the basis of a "meaningful projection from reliable data," in more than insignificant amounts. 613 F.2d at 955. The Court also held that FDA had discretion, even if a substance migrated in detectable amounts, to exempt it from the component element of the food additive definition if its presence could be classified as "de minimis" in terms of public health significance. 613 F.2d at 955. While the Court declined to elaborate on what constituted "significant" migration, it clearly recognizes that some level of migration is insignificant.

1979-1985. Despite the Monsanto decision, from 1979 to 1985, essentially no progress was made on the "no migration" question, although it was continually raised and discussed in a variety of circles. For example, from 1980 to 1984, an effort was undertaken by the leading trade groups and companies in the food industry to bring about comprehensive amendments to the Food, Drug and Cosmetic Act. On behalf of the packaging industries, The Society of the Plastics Industry participated in this work and encouraged the inclusion of language in the Bill to open the door for a return to rationality in packaging regulation. The proposed amendments were ultimately embodied in identical bills entitled "Food Safety Amendments of 1981." The bill was introduced as

S1442 by Senator Hatch in the Senate and by Congressman de la Garza as HR4014 in the House of Representatives. Introduction was in the Fall of 1981.

As a result of the introduction of this legislation, a Subcabinet Council Working Group was appointed by the government, which included top Agency officials who were delegated to consider the amendments and how they might be improved, implemented, or otherwise acted upon. Extensive presentations were made to this Subcabinet Council Working Group by representatives of the packaging industries, as well as many others. The efforts of the group and the legislation simply faded away so that it would appear this activity, like so many others before it to try to bring about a sense of balance in the food additive regulatory scheme, disappeared without anything definitive being done.

June, 1985. Dr. Sanford A. Miller, Director of the Center for Food Safety and Nutrition of FDA, spoke to the Food, Drug and Cosmetics Packaging Materials Committee of SPI at its Annual meeting. In his talk, Dr. Miller indicated a belief that the time might have come to work on a threshold of regulation concept for indirect additives.

September 27, 1985. On this date, a lengthy letter was submitted to Dr. Miller about a meeting on the threshold of regulation concept which was subsequently scheduled for November 25, 1985. At that meeting, a complete blueprint (including suggested new forms for a pre-marketing notification system) for action to reduce the paperwork — especially the time required for action on indirect additive matters — was advanced to the Agency. A promise was made that some definitive action would be taken on the threshold of regulation idea soon with the prediction of just a few months before something would be done.

December 3, 1985. At this time, a report was given at an SPI meeting regarding the discussions with the FDA about the concept of developing a scientific basis for a threshold of regulation. The aim was to work towards setting a base upon which FDA jurisdiction could be determined, so SPI pressed hard to cooperate with the Agency by advancing its ideas promptly.

1987. In this year, then Commissioner of the FDA Frank E. Young indicated at a meeting of the Technical Association for the Paper and Pulp Industry (TAPPI) on September 9, 1987, that the Agency was pushing the matter of developing a threshold of regulation position and would be making its ideas known shortly. It was well known by this time that Dr. Alan M. Rulis of the FDA staff had been put in charge of this activity, and, indeed, a seminal paper on this subject was prepared by Dr. Rulis and published in 1987. The paper, entitled "De Minimis and the Threshold of Regulation," was an excellent starting place and has since led to further work being done by the Canadian Centre for Toxicology operating under a grant provided by The Society of the Plastics Industry.

April 6, 1988. FDA issues a tentative final rule entitled "Colorants for Polymers," 53 Rd. Reg. 11402. Much to industry's dismay, the language of the Tentative Rule indicates that FDA is reverting to its pre-Monsanto position that all food contact substances may be expected to become components of food and are therefore food additives. The Agency's reasoning in the Preamble to the Tentative Rule, and especially its citation of Monsanto v. Kennedy, 613 F.2d 947 (1979) to support its position, turns Monsanto inside out. The proposal is reminiscent of John Harvey's 1961 "Catch 22" pronouncement that a packaging material is a food additive if one does migration studies to determine whether it is or not.

SPI protested vigorously to FDA regarding its return to its discredited interpretation. In response to industry criticisms, Gerad McCowin, Director of FDA's Division of Food and Color Additives, felt it necessary to clarify the Preamble in an April 19, 1988 speech concerning the Tentative Final Rule. With regard to the misunderstanding caused by the Preamble concerning the food additive status of colorants for which there is data showing no detectable migration, Mr. McCowin said:

> Some concern has been raised that the (Preamble's) response to the comment (that most colorants will not migrate and should not be regulated as food additives) can be read as implying that all colorants must be regulated as food additives . . . I don't believe that was our intent . . .There have been some occasions where we have received letters and even petitions for colorants that were accompanied by data showing no migration of the colorant at the level of analytical sensitivity. If the analytical sensitivity has been sufficiently low, we have written letters to companies advising that we did not believe their specific use of the colorants required regulation as a food additive. (Emphasis in original.)

Despite Mr. McCowin's assurances, the Preamble's statements that "colorants will migrate to food from all polymers" and "FDA has tentatively decided that colorants are food additives under the act" remain on the record. The rulemaking is still pending so it is hoped that the final record will be changed as Mr. McCowin has suggested.

March 29, 1990. This date marked the submission of the final report of the Canadian Centre for Toxicology's landmark study entitled "Evaluating Trivial (De Minimis) Carcinogenic and Non-carcinogenic Risks From Chemicals." This study provides a substantial new scientific basis for proceeding with the establishment of a proper threshold of regulation. It is hoped that it will inspire the FDA to do something definitive about this problem. In the letter of transmittal which accompanied the study, it is interesting to note that the Centre has reached the conclusion that "for substances lacking genotoxic potential, a regulatory threshold ... in the range between 5 and 10 ppb in the diet would be consistent with current FDA Redbook Guidelines." This conclusion indicates, as the Canadian Centre letter states in a footnote, that the dietary range mentioned is roughly equivalent to between 50 and 100 ppb extraction into the usual food simulating solvents. In short, the Canadian Centre Study supplies a new scientific support document for the old Ramsey Proposal introduced in 1969. Adoption of this proposal, even 21 years later, would be an enormous step forward. To do this, the Agency need only act upon Citizen's Petition 7CP3313 by adopting what it recommended or a reasonably modified version thereof.

Literature Cited

(1)   Harvey, "Food Additives and Regulations." 17 Food Drug Cosmetic Law Journal 275 (April 1962).

RECEIVED April 17, 1991

Chapter 9

Food-Contact Materials and Articles

Standards for Europe

J. F. Kay

Food Science Division I, Ministry of Agriculture, Fisheries, and Food, Ergon House, Nobel House, 17 Smith Square, London SW1P 3JR, United Kingdom

As part of its programme to create a Single European Market, the European Commission is harmonising rules governing materials and articles intended to come into contact with foodstuffs in order to remove technical barriers to trade and to protect the consumer. A framework Directive was adopted under which specific directives relating to particular food contact materials could be adopted. The key provisions of this Directive are that materials should be manufactured in accordance with good manufacturing practice and not endanger human health in use. Specific Directives already exist, e.g. on vinyl chloride monomer and regenerated cellulose film (cellophane). The recently adopted Plastics Directive lays down a positive list of monomers which can be used to manufacture food contact plastics. Future directives will include paper and board, already at an advanced stage of consideration by the Council of Europe (a non EC body), plus further plastics work. There is an active co-ordinated surveillance programme of work in the U.K. providing scientific data to assist the above bodies in proposals for future legislation.

It is important to remember when considering legislation in the food and food packaging sectors what its primary aims are. In the food packaging industry it can generally be said that it is intended to ensure that the packaging results in food being safe, wholesome and of the nature, substance and quality demanded by the purchaser. The maintenance of a safe and nutritious food supply has long been regarded as a matter of the utmost importance in the United Kingdom, as in many other countries. Indeed, the first Sale of Food and Drugs Act was enacted well over 100 years ago. Since that time legislation on food has continually been updated to take account of new developments.

0097–6156/91/0473–0104$06.00/0

To ensure that food safety legislation affords adequate and effective consumer protection, legislation must be based on sound, scientific principles.   This is the UK Government philosophy and in consequence an extensive advisory system has been established utilising extensive impartial, independent, scientific and medical opinion on all safety-related aspects of food, including food packaging. This system is increasingly providing data for use in negotiations in the European Community.

Controls on Food Packaging

The principle reasons for the use of food packaging are to allow for the hygienic handling of food, prevent microbial contamination and more recently through modification of the atmosphere around the contained food, to extend the shelf-life.   However, the package itself can represent a source of chemical contamination through the transfer, or more accurately migration, of substances from the packaging material into the food. Many regulatory authorities around the world have recognised that it is necessary to control this migration and have enacted extensive legislation.   Much of this legislation includes positive lists of permitted substances together with appropriate restrictions.   A limit on the total migration of the constituents of the material, known as the global or overall migration, may also be included.

EC Legislation on Food Contact Materials

As part of its programme to create a Single European Market, the EC is drawing up rules governing materials and articles intended to come into contact with foodstuffs in order both to remove technical barriers to trade and to protect the consumer. The starting point was to establish a framework directive under which specific directives relating to particular food contact materials could be adopted. The key provision of the materials and articles framework Directive 89/109/EEC and of UK Regulations implementing it (The Materials and Articles in Contact with Food Regulations 1987 [S.I. 1987 No. 1523]) is that materials and articles shall be manufactured in accordance with good manufacturing practice, i.e. in such a way that under normal or foreseeable conditions of use they do not transfer their constituents to foods with which they are, or are likely to be, in contact, in quantities which could:

(i) endanger human health, or
(ii) bring about a deterioration in the organoleptic characteristics of such food or an unacceptable change in its nature, substance or quality.

In order to reinforce its general provisions, the framework Directive provides for specific directives to be developed for individual food contact materials, including paper and board, and lists the various measures that these directives may include. Directives have already been agreed which lay down limits on the release of lead and cadmium from ceramic ware (84/500/EEC), establish a positive list of substances that may be used in the production of regenerated cellulose film (83/229/EEC); control the migration of glycols from such film (86/338/EEC), and specify the symbols which may be used to accompany food contact materials and articles (80/590/EEC). A series of Directives controlling residues of vinyl chloride monomer have also been adopted (78/142/EEC; 80/766/EEC; 81/432/EEC). These Directives have either been implemented in UK law or are being so implemented. Two Directives lay down the basic rules for testing migration from plastics (82/711/EEC) and, for plastics intended for contact with specific foodstuffs or groups of foodstuffs, specify which simulants (food simulating liquids) are to be used in each case (85/572/EEC). They are required to be enacted in UK law at the same time as the first Community positive list.

EC Commission Directive 90/128/EEC on food contact plastics comprises a limit on overall migration and a positive list of monomers and other starting substances. This was adopted earlier this year. Under this Directive plastics materials and articles shall not transfer their constituents to foodstuffs in quantities exceeding 10 mg/dm² of surface area of the material or article (overall migration limit). However this limit will be 60 mg/kg of foodstuff in certain cases.

The second aspect of the Directive comprises a positive list of monomers and other starting substances. All of the monomers and other starting substances have been evaluated by the Commission's toxicological advisory committee, the Scientific Committee for Food (SCF). Substances which exhibit some toxicity but which are essential for polymer manufacture, are subject to stringent limits, either in the plastic or in the food and such limits incorporate a large safety factor reducing further the level which would be acceptable on toxicological grounds in the worst migration situation.

Substances for which there were insufficient toxicological or technological data to enable a judgement to be expressed will be re-examined by the SCF in a rolling programme up to 1992. Any substance that cannot be evaluated favourably will be deleted from the Directive, apart from those substances subject to long term testing, for which a deadline of 1996 has been set.

The UK approach

It has long been recognised in the UK that one of the principle protections for the consumer from any hazard arising from the presence of contaminants in food is the existence of a vigilant and responsible industry, and its willingness to co-operate with Government in this area. As part of this process of co-operation, it has been usual to involve the food and food packaging industries in any official information-gathering exercises, designed to assess the levels of contamination of food which might arise from the use of food contact materials. Such exercises have normally been undertaken as part of MAFF's food surveillance programme and overseen by a senior UK Government interdepartmental advisory body, the Steering Group on Food Surveillance (SGFS). This body has a remit to keep under review the possibilities of contamination of any part of the UK food supply and to review where necessary the intake of individual additives and nutrients. Recommendations are made to Ministers responsible for food quality and safety on the programme of work necessary to ensure that the food intake of the population is both safe and nutritious. The work of this Steering Group will continue primarily to provide new information which can be fed back into the statutory control system to produce modified or new legislation as necessary.

Following the carrying out of a number of successful ad-hoc exercises in the 1970s on monomers of toxicological concern, the SGFS decided that it would be appropriate to establish a working party to keep under review the possibility of contamination of any part of the UK food supply by chemicals arising from food contact materials. Thus the Working Party on Chemical Contaminants from Food Contact Materials was established during 1984, with a membership drawn largely from manufacturers and users of such materials.

In its first year the Working Party tackled a number of matters of immediate importance, notably plasticisers. A Sub-group was set up to undertake an in-depth examination of these materials and its findings have been published as Food Surveillance Paper No. 21 (1). One particular aspect of this Report dealt with 'cling film' and the migration of its plasticisers, especially during microwave cooking.

The work of this Sub-group, together with toxicological evaluation, enabled guidance to be given to the public on the use of such cling-wrap materials particularly in microwave cooking and for the industry to use this as a platform to develop new materials with a lower migration potential. A second report on this topic will be published later this year. Other Sub-groups have been established by the Working Party to undertake in-depth reviews of colourants and aids to polymerisation with the objective of drawing up recommendations on the usage of such substances. Having now successfully completed their programmes of work, the Sub-groups on Plasticisers and also Aids to Polymerisation have been disbanded. Another Sub-group is carrying out a survey on the uses of food contact paper and board with the intention of calculating dietary intakes of any components of paper and board that migrate into foodstuffs. The presence of dioxins in paper and board and in food in contact with these materials has been investigated by this group and does not appear to pose a problem as levels are very low. A Food Surveillance Paper similar to that produced on plasticisers is in preparation on this topic. Preliminary data on dioxin levels migrating from food contact paper and board have been published in Department of the Environment Pollution Paper No. 27 (2).

A fifth Sub-group was established with a remit to consider food contact materials in toto and elaborate a coherent strategy for their systematic examination. This Sub-group has completed its study and its report has been included in Food Surveillance Paper No. 26 (3). An important aspect of all of the Working Party's work is the commissioning of an appropriate research programme both within and outside the Ministry's own Food Science Laboratory. In order to produce a research programme with commonly agreed priorities, a Sub-group on Research Priorities was established to oversee MAFF's research programme on food contact materials. The research undertaken by the Working Party between 1984 and 1988 is also detailed in Food Surveillance Paper 26 (3).

Research programme

The Ministry's own Food Science Laboratory has carried out research on food contact materials since the mid-1970s. This has involved both the provision of surveillance data on monomer levels and also investigating more fundamental potential problem areas. These studies have been conducted in close collaboration with the plastics industry and initially established the chemical identities of the full range of ingredients of commercial PVC formulations.

In recent years work has been carried out in support of various Sub-groups or in collaboration with outside contractors. Many data have recently been generated for the forthcoming Second Report of the Sub-group on Plasticisers. Methods have been developed to measure epoxidised soya bean oil (ESBO) and polymeric plasticisers in both plastics and foods. Phthalates are used in printing inks and their migration into foods have been determined.

The effect of aseptic processing on migration has been studied. Both overall migration and the migration of a specific antioxidant from polypropylene pots treated with a commercial strength hydrogen peroxide solution have been measured. Antioxidant migration from the pots does not appear to depend on treatment.

Microwave oven usage has become increasingly popular in the UK over the past few years and this has resulted in a wide range of plastics cookware being specifically produced for use in these ovens. The Food Science Laboratory has set up a programme to examine the materials used to fabricate these cookware items and initially has concentrated upon poly(vinylidene chloride) film and polyethylene terephthalate trays. Since existing EC-laid down testing conditions do not refer to microwave usage, a protocol for testing materials intended for use in microwave ovens is also being established. Susceptor materials interact with microwave energy to achieve high temperatures very rapidly and are intended for browning and crisping applications such as cooking potato chips and pastries. Studies have shown that the migration of PET oligomers from susceptor pads is low (4). Total levels of migration of PET oligomers were found to range from 0.02 to 2.73 mg/kg, depending on the foodstuff and the temperature attained during cooking. Further studies are being conducted into the potential migrants from the adhesive and pulpboard layers of these materials.

Studies on repeat-use articles have resulted in a change of EC proposals for migration testing of these articles. The third test is now applicable for all materials intended for repeat-use unless there is evidence that migration does not increase with successive exposures.

Analysts will require standard samples of plastics and plastics additives to enable the proposed EC legislation on food contact plastics to be enforced. The feasibility of establishing a reference collection of standard samples is being explored.

The UK government is currently undertaking a review on the uses of mineral hydrocarbons in cheese waxes and food contact materials. Migration of these substances into cheeses and into skinless sausages (as a result of their use on sausage casings) has been studied.

The feasibility of producing plastics extracts under conditions representative of normal usage is under consideration. This would enable toxicological testing to be carried out not just on individual components, but on the total extract of the plastic, including transformation products.

Since the early 1980s, projects have been placed with outside contractors to supplement those being carried out in the Food Science Laboratory. Although a consolidated research programme could not be established immediately, this was seen as the ultimate objective and the last few years have seen considerable progress towards this goal. Some of these studies are described below.

Assessment of consumer exposure to the differing types of primary (retail) packaging. The objective of this study was to determine the surface area of each type of packaging material coming into contact with the 'average' diet. This information will prove invaluable in the estimation of intakes of migrants. In the future this project will be extended to cover caps and closures and beverages.

Mathematical and physical models of migration. The various mathematical and physical models that have been developed around the world to predict migration have been critically reviewed. In the second phase of this work, the long-term objective is to develop a limited number of models which will cover the behaviour of the majority of cases of practical interest, leading to a basis for legislation regarding migration limits.

The release of contaminants by leaching from 'boil-in' bags. Following initial studies on the effect of polymer morphology on diffusion, solubility and permeability co-efficients, this project concentrated upon quantifying substances migrating from 'boil-in' bags under natural usage conditions. The various layers of commercial bags were examined individually starting with the polyolefin food contact layer.

The effect of gamma- and electron-beam irradiation on food contact plastics. The effect of gamma-irradiation on additives present in food contact plastics has been examined. Initially organotin stabilisers were studied followed by phenolic and phosphite antioxidants. A study is now being undertaken to determine the effects of electron-beam irradiation on such additives.

Thermal stabilities of polymers used in food contact plastics. Substances migrating from 'ovenable' plastics under actual usage conditions were quantified with the emphasis being placed on polyethylene terephthalate trays.

Examination of the possibility of setting testing priorities for substances used in food contact materials. The correlation between chemical structure and toxicity is being investigated for substances on the draft EC positive lists of monomers and plastics additives. Quantitative Structure Activity Relationships (QSARs) are being developed for such substances, with the aim of prioritising these substances for further study.

The research programme is being widened to include the study of

migration from food contact materials at elevated temperatures and migration from metal containers.
Liaison between contractors is strongly encouraged in order to share expertise and avoid duplication of effort. An annual meeting is held at which contractors summarise progress during the preceding year. In order to ensure that the work is commercially relevant, an industrial rapporteur is appointed to each project to assist the Ministry's project officer.
These rapporteurs have appropriate specialist or industrial knowledge and are involved with a project at all stages from finalising the objectives and identifying potential contractors through to recommending further work. To date this system has proved highly successful with contractors generally reacting very favourably towards it.

Conclusions

Up until now, the UK has opted for a basic statutory framework combined with a monitoring programme carried out with the support of industry. In the next few years, more extensive legislation will be adopted as part of the EC's harmonisation programme. It is essential that such legislation is based on good science and the UK will continue its extensive surveillance and research programme to achieve this. The implications of forthcoming EC legislation are being discussed with Industry.

Literature Cited

1   Ministry of Agriculture, Fisheries and Food; Survey of plasticiser levels in food contact materials and in foods; Food Surveillance Paper No. 21; HMSO: London, UK, 1987.
2   Department of the Environment; Dioxins in the Environment; Pollution Paper No. 27; HMSO: London, UK, 1989.
3   Ministry of Agriculture, Fisheries and Food; Migration of substances from food contact materials into food; Food Surveillance Paper No. 26; HMSO: London, UK, 1989.
4   Castle, L., Mayo, A.K., Crews, C. and Gilbert, J. J. Food Protect. **1989**, 52, 337-342.

RECEIVED April 25, 1991

Chapter 10

Food–Package Interaction Safety

European Views

K. D. Woods

Waddingtons Cartons Limited, Leeds LS10 3TP, England

Waddingtons Cartons Limited is the holder in the UK, France and Germany of the Sieferth microwave susceptor patent, and as such this is where our main interest lies in the area of food package interactions. As a commercial producer and supplier of susceptor laminate packaging our interest in the safety of the material can be divided into three categories. Firstly, as a responsible company, we are concerned about the health and safety of the public and we therefore needed to assure ourselves that the susceptor material which we supply to our customers possess no threat to the public's health. Secondly we are required to conform to the regulations in force in the countries to which we supply. This is somewhat complicated since most of the food packaging regulations were drawn up some time ago and did not foresee susceptor materials, and the associated high temperatures which can be created in the packaging. Finally, we needed to guard against bad publicity which may severely affect the sale of susceptor packaging and the products which are packed in them.

Early on there was much speculation on the safety aspect of susceptors because of the relatively high temperatures that are reached under normal conditions of use (i.e., up to 220°C/450°F). This fear has largely died away as an increasing number of investigations have been carried out. Our own susceptor was tested by several people including James River in the USA (with whom we have a technical know how agreement) and the Ministry of Agriculture, Fisheries and Food (MAFF) in the UK. None of the studies indicated that there was any cause for concern.

In the mean time much is being done by the European regulatory bodies to produce suitable legislation to cover susceptors. We are cooperating in this and have become involved in some of the work, but we can only wait to see what these regulations will be. When the legislation is produced we expect that our material will conform.

0097–6156/91/0473–0111$06.00/0
© 1991 American Chemical Society

Currently most of the safety work has looked at the susceptor on its own. In the UK many of the susceptor packs on sale are heated in the microwave oven after first removing all additional packaging and therefore require no additional investigation into its safety. There is though a requirement for greater convenience for the food packer and the consumer, together with a wish to keep costs to a minimum. These factors have all led to a demand for the susceptor to be incorporated into other packaging components Typically, the route employed is to window patch a piece of susceptor paper into a paperboard carton. Window patching involves gluing the susceptor to the carton during the carton manufacturing process. This uses a machine traditionally employed to stick clear film over apertures in the carton, so producing a decorative window in the carton, hence the term window patching.

The integration of the susceptor into the pack can potentially subject the various components in the pack to the high temperatures generated by the susceptor. Normally these components include window patch adhesive, cartonboard and inks. Other adhesives used to hold the carton together may also be exposed to higher than normal temperatures, although since these tend not to be in direct contact with the susceptor, the temperatures will be significantly lower than the 200°C produced by the susceptor. While the chances of any undesirable substances being produced and transferring to the food is low, this area needs to be investigated.

Special window patch adhesive have been developed and tested by a number or adhesive manufacturers Less development has been undertaken on the inks. Many of the inks used for non-microwave applications can be shown to contain organic pigments which in contact with susceptors in the microwave oven sublime. While the ovens are well ventilated and the small quantity of ink is on the outside of the pack it is unlikely that any of the pigment will transfer to the food These pigments are easily avoided by specifying non-sublimable pigments to the ink manufacturer, so removing the potential risk. Other volatiles may also be given off from the inks and these will depend upon their type and formulation.

The choice of board used for the carton also presents additional problems. Carton boards have a surface coating to give a good smooth surface suitable for printing. The formulations of these are closely guarded by the board mills since the coating performance is one of the main differentiations between alternative boards. Susceptors, since they are usually unprinted, are produced using uncoated boards, but in printed cartons the coatings are another unknown commodity at susceptor temperatures.

In the USA only two types of cartonboard are generally available, Solid White Board and Chipboard Chipboard is produced from waste paper and is clearly unsuitable for use in contact with susceptors, since no guarantees can be given as to what substances may be contained in the finished board. In the UK and Europe the choice is much greater with the availability of Folding Box Boards (FBB) which are significantly cheaper than equivalent Solid White Boards.

FBBs are produced from one or more types of pulp. The pulps can be split into four types, Mechanical, Thermo-Mechanical (TMP), Chemo-Thermo-Mechanical (CTMP) and Solid Bleached Sulphate (SBS) pulps. SBS is effectively pure cellulose fibre, all other components of the wood (about 50%) having been removed in the pulping process. The other pulping processes remove less of the wood's non-cellulose components, such as lignin, resin acids, fatty acids, turpinoid compounds and alcohols. Typical yields from these pulping processes vary from 65% to 95%. Again these additional compounds pose a potential risk.

Since all the boards available vary because of different coatings and different combinations of pulps, we consider it wise to test each board we wish to use if it has not previously been analysed.

While there is probably little risk from all these potential problems we consider that all integral susceptor packs should be carefully considered, and any analytical work thought necessary carried out before the product is placed on the market. Below is a brief description of some of work that has been carried out on our packs.

Research Overview and Results

In the first study the particular pack of interest was a Microwave Chips (French Fries) carton. This consisted of a top load carton printed with UV inks and varnish, with a rectangle of susceptor window patched on to the base of the carton. After filling the carton with chips, the carton is closed with side seams of water-based and hot-melt adhesives, located approximately a third of the way up the side walls.

To answer some of the questions about the safety of the materials, work was commissioned to be carried out at the Leatherhead Food Research Association. The work was funded by the food manufacturer, LMG Mardon (a susceptor licensee in the UK) and Waddingtons Cartons Limited.

The work was carried out in two parts. First the likely volatiles were analysed in a simplified system under "worst case" conditions. The use of a semi-solid fat simulant (see later) gave slightly higher susceptor temperatures than with actual food and so satisfied the "worst case" criterion required. When these compounds had been identified, actual product, which had been microwave heated in the complete carton was tested to see if these compounds did in fact migrate into the chips.

Temperature measurements at the glue seams had recorded temperatures of between 70°C and 112°C, but measurements of the internal surface of the side walls had shown a maximum temperature of 131°C. Therefore in order to be certain that we had a worse case situation a temperature of 140°C was considered to be the temperature suitable for analysis of compounds which may arise from the adhesives during heating and may then transfer to the food product.

The volatile compounds given off by the glues were analysed using gas chromatograph/mass spectrometry (GC/MS). Although difficult to identify precisely from their mass spectra many of the peaks from the hot melt adhesive appeared to be of natural origin and typical of compounds which would be given off by plant extracts. The resin used in the hot melt adhesive is a derivative of limonene, the main hydrocarbon fraction of many essential oils, and is the likely source of these compounds.

The other components of the packaging were tested by microwave heating in a sealed vessel together with a semi-solid fat simulant which had been developed at the Ministry of Agriculture, Fisheries of Food (MAFF) in the UK. The simulant was then extracted using Lichens-Nickerson apparatus, which gives good recovery of compounds with a volatility range suitable for GC. The resultant extract was then analysed by GC/MS.

Earlier work with the MAFF semi-solid fat simulant had shown good correlation between the susceptors temperature profile in contact with 100g of simulant (in this carton) and with 100g of chips (both in a 650W oven with neither in a sealed vessel). The simulant temperature were slightly higher. These temperatures were recorded with an earlier formulation simulant than that actually used in this test, but the temperatures obtained were not expected to be radically different.

The simulant used consisted of 35% water, 25% olive oil and 40% diatomaceous earth, i.e., Celite. The trials were carried out in a 600W oven. 100g of the simulant together with the test sample were heated for 4 minutes on full power in a sealed glass vessel just large enough for the carton to be placed inside.

In order to obtain some idea of where any volatile components detected had originated from, four tests were carried out. These were:-

a) Blank - 100g of simulant inside glass vessel with susceptor patch taped to outside.

b) Susceptor only - susceptor patch inside vessel, simulant packed around it (50g above, 50g below).

c) Unprinted carton - including susceptor and window patched adhesive. 100g of simulant inside carton.

d) Printed carton - as c) above but carton printed with UV inks and varnish.

Before the simulant was extracted, 1ppm of deuterated benzene and naphthalene were added as standards. GC/MS analysis showed some increase in concentration of compounds found in the blank together with the appearance of a number of additional compounds. The concentration of these compounds was estimated to be in the range of 0.1 to 5ppm.

Compounds with increased concentration are listed below:-

Heptane	Octanal
Pentan-1-ol	Propyl Furan
Octane	Octenal Isomer
Hexanol	Octenol Isomers (2)
Heptanal	Nonanal
Heptenal	Nonenal Isomer
Alkenes (3)	Decenal Isomers (3)
Alkane	Decadienal Isomer (2)
Methyl Furfural	Alkyl Cyclohexenol ?
Alkyl Furan	Alkyl Cyclohexenone ?

Additional compounds identified were:-

Furfural	Susceptor, Plain and Printed carton
Decatrienal Isomer?	Susceptor, Plain and Printed carton
$C_{15}H_{24}$ Sesquiterpene	Plain and Printed carton
Benzophenone	Printed Carton only

Benzophenone is a photoinitiator used in UV inks and clearly this is where it originated.

Having identified the above compounds which could potentially be transferred to the food, three duplicate tests were carried out on the actual product to establish if any of these compounds did in fact transfer to the chips.

The complete packs, including 100g of chips were cooked in a 650W oven according to the heating instructions given for the product, i.e., 3 minutes at full power followed by 1 minute standing. The french fries were then spiked with 1ppm deuterated naphthalene and extracted using Lichens-Nickerson apparatus into diethyl ether. This extract was then analysed as for the simulant extracts, by full scan GC/MS. Many compounds seen in the simulant experiment were found in the extracts from the chips, but at a much lower level. The concentration of these compounds were all estimated to be below 1ppm.

Of the additional compound not found in the simulant blank but detected in the other simulant tests only the decatrienal isomer and benzophenone were found in the chip extract. This suggests that furfural and the sesquiterpene, which were not present in the chip extract, had been derived from the olive oil used in the simulant (and therefore was not considered a cause for concern).

The concentration of benzophenone was calculated by comparison with a standard of authentic compound. The benzophenone level found (0.2 - 0.3 ppm) is below the limit for specific migration proposed by European Community regulations, and as such should not present a problem. The concentration of decatrienal isomer was also below 1ppm.

These results on the actual pack support the view that the simulant represented a "worst case" compared to actual use.

None of the components given off from the glues were found in the real food analysis, therefore, these adhesives can be considered safe for non-direct contact with the food when positioned away from the susceptor.

While a total assurance on safety is not possible, given that the experiments carried out can only detect compounds which are amenable to analysis by GC/MS and at around the ppm level, the test results found nothing to worry about from a safety point of view.

A separate study looked at a cartonboard proposed for use with a window patched carton.

The proposed cartonboard was a folding box board (FBB) consisting of five plies. The back ply is mainly mechanical pulp with a small proportion of chemical pulp. The three middle plys are produced from mechanical pulp with the top ply being produced from chemical pulp. This top surface is then coated to give a good surface for printing.

The susceptor pack incorporating this board was then tested at our customer's research centre using the following method:-

A 5cm x 6cm piece of the susceptor laminate / window patch adhesive / FBB placed in a sealed glass vial was heated in a 600W microwave oven for 10 minutes. The headspace in the vial was then analysed using GC/MS.

The conditions chosen (i.e., 10 minutes in a 600W oven without any additional load) are based on the time required for the product to be heated. It is recognised that these conditions are far more severe than in practice where the food load is present, but these more severe test conditions were considered valid for the purpose of the tests, that is to determine if the pack is safe even in the worst case situation.

The test on the susceptor paper laminate / window patch adhesive / cartonboard composite showed that benzene was generated at levels of 0.05 μg/cm². The benzene could have originated from either the susceptor laminate, the window patch adhesive or the folding box board. Further tests were therefore initiated to determine where this benzene had in fact come from.

A large quantity of work had been carried out previously on the susceptor laminate. None of this had shown benzene could be produced on heating. This suggested that the susceptor was not the source. Testing the susceptor laminate alone under the same conditions confirmed this.

The window patch adhesive was next to be ruled out. 5cm x 6cm pieces of glued and non-glued susceptor / FBB were tested. These both produced similar levels of benzene which showed this was not the source.

These results pointed strongly to the board being the source of the benzene. Since without the susceptor, the FBB alone will not heat to the required temperatures in a microwave oven, the above test conditions could not be applied. To confirm that the FBB was indeed the source a vial containing a 5cm x 6cm sample of the FBB alone was heated in a muffle furnace to 200°C for 2 minutes and the resulting headspace analysed. The presence of benzene confirmed that the benzene was a product of the pyrolysis of the FBB. The results of the benzene determination are listed in table 1.

Table 1. Benzene levels generated by package components

Susceptor/window adhesive/FBB	0.05 μg/cm²
Susceptor/90% window adhesive/FBB	0.07 μg/cm²
Susceptor laminate/FBB	0.06 μg/cm²
Susceptor laminate only	ND (<0.003 μg/cm²)
FBB only (heated to 200°C)	0.01 μg/cm²

It should be noted that the level of benzene (0.01 μg/cm²) produced when FBB only was heated to 200°C is much lower than that found in the microwave tests (0.05 - 0.07 μg/cm²). The lower figure is likely to be a more accurate representation of the amount of benzene which will be produced from the carton board, since the temperatures the sample has been exposed to will be closer to those experienced under actual conditions of use. The sample heated in the vial is likely to reach much higher temperatures due to the low load and the multiple layers of susceptor (subsequent tests have indicated that temperatures up to 270°C are possible).

The levels of benzene produced in the microwave tests, which would equate to approximately 30ppb if all the benzene were to migrate to the food, were not considered to pose any health threat since work carried out at the Ministry of Agriculture, Fisheries and Food show that due to its high volatility, little if any of the available benzene would be absorbed by the food. Even so, due to benzene in food being such an emotive subject with the general public, the decision was made not to use this or similar boards for cartons including susceptors.

While it has not been confirmed, we believe that either lignin or compounds in the coating of the FBB when subjected to the high temperatures created by susceptors produces benzene as one of the pyrolysis products.

The FBB used in the french fries pack described earlier, which is produced mainly from Chemo-Thermo-Mechanical pulp (CTMP) was found to be free from benzene, so not all FBBs are suspect. We are currently having tests carried out on boards produced from the different pulps and by different board mills to find suitable low cost alternatives to SBS boards which are safe and can be used without causing undue concern amongst the press and general public. Hopefully, this work will also shed some light on the source of the benzene found.

The work we have been involved in clearly shows that if proper care is taken there are no safety problems with susceptors inside cartons so long as suitable components are used.

RECEIVED March 1, 1991

Chapter 11

Influence of Microwave Heating on the Formation of N-Nitrosamines in Bacon

Shaun Chenghsiung Chen[1], Bruce R. Harte[1], J. Ian Gray[2], and Alden M. Booren[2]

[1]School of Packaging and [2]Department of Food Science and Human Nutrition, Michigan State University, East Lansing, MI 48824

Bacon was cooked in a teflon-coated frying pan and in transparent and susceptor packages in a 700 W microwave oven. Bacon slices were prepared to three "degrees of doneness": "undercooked", "overcooked" and "similar" to standard pan frying (3 minutes on each side at 340°F) by controlling the time in the microwave oven. The package interfacial temperature and bacon temperature were monitored using a Luxtron fluoroptic probe system. Final bacon temperatures of 167°C (330°F) and 143°C (290°F) were observed for samples cooked to the "same degree of doneness" as fried bacon in the susceptor (2.5 minutes) and transparent (3 minutes) packages, respectively. Compared to frying, smaller concentrations of N-nitrosopyrrolidine (NPYR) and N-nitrosodimethylamine (NDMA) were present in bacon cooked either in the transparent or susceptor packages. Greater concentrations of NPYR were found in bacon cooked on susceptors in comparison to cooking in a transparent package. Formation of NPYR in pan-fried bacon was correlated with the concentration of nitrite in the raw product.

Convenience in food preparation makes the microwave oven a necessary component in today's households. With microwave ovens present in approximately 80% of U.S. homes and retail sales of microwave foods at several billion dollars, the development of microwave foods has become an increasingly important research and development focus for many food companies (1,4). The time savings involved in "in-package" cooking is a significant convenience feature for many working people.

0097–6156/91/0473–0118$06.00/0

Insufficient crisping and browning have been the most common negatives associated with microwave cooking (*2*). Packaging techniques which enhance heating to cause dehydration and browning on the surface of products during microwave cooking are currently being utilized. These energy-absorbing packaging materials, called susceptors, utilize controlled thicknesses of metal vacuum deposited onto paper, paperboard or film carriers. The metal droplets absorb microwave energy and generate localized heating (*2-4*). Temperatures observed on the surface of the susceptor have been reported in excess of 200°C (*5-7*). At these temperatures (>200°C), several safety concerns have been raised including potential migration of package components into food products (*6-8*). It is also possible that the higher temperatures involved may enhance the formation of N-nitrosamines in microwave-cooked bacon.

The formation of N-nitroso compounds in cooked bacon is affected by many factors (*9*). These include: 1. Cooking method - using a preheated frying pan was found to produce more N-nitrosamines than cooking in an initially cold frying pan (*10,11*). 2. Cooking time and temperature - increasing cooking time and/or temperature can increase N-nitrosamine formation in cooked bacon (*10,12*). 3. Lean-to adipose tissue ratio - since the formation of N-nitrosopyrrolidine (NPYR) appears to be primarily associated with the adipose tissue, a high adipose-to-lean ratio can enhance N-nitrosamine formation in cooked bacon (*13*). 4. Inhibitors - ascorbic acid, ascorbyl palmitates and α-tocopherol act as "blocking agents" and compete with secondary amines for the nitrosating species, thus less nitrite is available for nitrosation (*14,15*). 5. Storage period - free proline can be formed by proteolysis during storage prior to cooking, and this can result in greater concentrations of N-nitrosamines after cooking (*16*). 6. Concentration of nitrite - since the formation of N-nitrosamines depends on the reaction of proline and nitrite, an increased concentration of nitrite induces greater formation of N-PYR (*17*). 7. Thickness of slices - slice thickness can influence penetration of heat into bacon, which can affect the formation of N-nitrosamines in bacon (*18*). 8. Liquid smoke - significant reduction of NPYR in fried bacon was found when liquid smoke was incorporated into the pork bellies in the curing brine.

Use of susceptors in microwave "in-package" cooking may increase the temperature at the surface of the bacon sample while reducing cooking time, both of which may affect the formation of N-nitrosamines.

The objectives of this study were to:

1. Measure the temperature of the product and package as a function of microwave cooking in susceptor and non-susceptor (transparent) packages.
2. Develop a correlation between product temperature and formation of N-nitrosamines in microwave cooking in comparison to frying.
3. Determine the effect of product composition on the formation of N-nitrosamines as a result of conventional and microwave cooking.

Methods and Materials

Preparation of Bacon Samples. Pork bellies were processed into bacon within 48 hours of slaughter in the Meat Laboratory, Michigan State University. They were injected using a multiple needle system to 110% of their green weight using a brine, to obtain a target concentration of 1.5% sodium chloride, 0.5% sucrose, 0.35% sodium tripolyphosphate, 550 mg/kg sodium ascorbate and 120 mg/kg sodium nitrite. Pumped bellies were placed in plastic bags and allowed to equilibrate overnight at 2°C. Bellies were transferred to an Elec-Trol laboratory smokehouse (Drying System Inc., Chicago, IL) showered with cold water for one min and smoked at 58°C (dry bulb temperature) for 4 hours, followed by three hours at 52°C (dry bulb temperature). The bacon was transferred to a cooler at 2°C overnight prior to slicing and packaging *(20)*.

To study the effect of nitrite concentration on the formation of N-nitrosamines, pork bellies were injected to 110% of their green weight to obtain a target concentration of 1.5% sodium chloride, 0.5% sucrose, 0.35% sodium tripolyphosphate, 550 mg/kg sodium ascorbate and 200 mg/kg nitrite and processed into bacon as described previously.

After smoking, the pork bellies were cut into 0.32 cm (1/8 in) thick slices. Bacon slices were placed on polystyrene trays, deposited into polyethylene pouches, and sealed under vacuum. The packaged samples were stored for seven days at 2°C prior to cooking. Cooked bacon samples were stored in polyethylene pouches (18 oz. Whirl-Pak, NASCO, Ft. Atkinson, WI), and placed in a freezer prior to analysis.

Two commercial slab bacons, purchased from a local retail store, were sliced (0.32 cm thick) and vacuum packaged in polyethylene pouches immediately after purchasing. Samples were stored at 2°C for seven days prior to cooking.

Determination of Residual Nitrite. Residual sodium nitrite ($NaNO_2$) content was determined in a 5 g sample of raw bacon using the AOAC procedure *(21)*. Quantitation was based on a standard curve plotting absorbancy at 540 nm versus concentration. The standard curve was made using 1, 2, 5, 10, 20, 30 and 40 ml of a 1 ppm (μg/ml) $NaNO_2$ solution. The amount of residual nitrite was determined by comparing absorbancy at 540 nm, and calculating relative concentrations from the standard curve.

Thermal Processing of the Bacon Samples. Bacon samples were heated conventionally in a frying pan and in two different microwave packages, to determine the effect of cooking method on the formation of N-nitrosamines. Frying was accomplished using a preheated teflon-coated electric pan set at 171°C (340°F) for 3 minutes on each side (6 minutes total). Three slices of bacon (80 - 100 g) were fried simultaneously. After frying, the fried bacon samples were placed on paper towels to permit removal of excess cooked-out fat, and then stored at -20°C until analyzed.

Microwave "in-package" cooking was achieved using a susceptor (aluminum metallized polyester, adhesive laminated to paperboard) and a microwave transparent tray (polyester coated paperboard) in a 700 watt microwave (model RS458P, Amana Refrigeration Inc., Amana, IA). The output of the microwave oven was determined prior to use, and found to be 680 watts.

The bacon samples were cooked in the microwave oven at full power. Three slices of bacon (80 - 100 g) were used in both microwave cooking methods. Cook-out-fat was drained off on paper towels. Heating times were varied to obtain a specific "degree of doneness". Cooking times of 2, 3 and 4 minutes were used for transparent in-package cooking, and 2, 2.5 and 3 minutes for the susceptor cooked bacon. Bacon prepared in the transparent package for 3 minutes or in susceptor packages for 2.5 minutes was observed by the authors to have the same degree of doneness as fried.

Determination of Temperatures During Heating. A thermocouple (Omega Engineering Inc., Gardiner, NY) was used to measure the temperature of bacon during frying. Four probes were placed at the interface between the bacon sample and the frying pan. Temperature readings were recorded every 30 seconds.

For bacon cooked in the microwave oven, package material and bacon surface temperatures were measured using a Luxtron 755 Multichannel Fluoroptic Thermometer (Luxtron, Mountain View, CA). Two MIW probes were placed onto the surface of the empty microwavable packaging material and/or interface between the samples and the packaging material. Temperatures were measured and recorded every 10 seconds using a data collection computer system.

Determination of N-Nitrosamines in Cooked Bacon. The amount of N-nitrosamine formed in cooked bacon was quantified using a dry column - thermal energy analyzer method (*21*). The dry column method which was used to extract N-nitrosamines from cooked bacon utilized dichloromethane ($CH_2 Cl_2$) and N-pentane as solvents.

An aliquot (5 μl) of the concentrated extract was injected into a GC-TEA (Varian Aerograph gas chromatograph, Model 3700, interfaced with a TEA, Model 502). A 2.7 m x 3.2 mm (ID) glass column (Supelco, Bellefonte, CA) packed with 15% Carbowax 20M-TPA was used for the separation of the N-nitroso compounds. Conditions were set initially at 140°C, then temperature programmed to 180°C at 7°C/minute. Peak areas for the N-nitrosamines detected were calculated using a Hewlett Packard Model 3390A Integrator (Walnut Creek, CA).

The analytical efficiency was monitored by determining the recovery of the internal standard, N-nitrosoazetidine (NAZET). Specific N-nitrosamines were identified by comparing their retention times to those of standard N-nitrosamines. Mass spectrometric confirmation of the N-nitrosamines was not attempted in this study and hence the N-nitrosamines can only be referred to as apparent N-nitrosamines. However, previous studies in our laboratories have utilized mass spectrometric techniques for the identification of these N-nitrosamines. The amount of apparent N-nitrosamines in the bacon were determined using their relative response area. N-nitrosamine concentrations in the cooked bacon were calculated

by: $N\text{-}nitrosamine \ (\mu g/kg) = (\dfrac{A_{sample}}{A_{standard}}) \ x \ (\dfrac{CONC. \ OF \ STANDARD}{wt \ of \ sample}) \ x \ 1000$ where,

A_{sample}, represents the response area of N-nitrosamines in cooked bacon. $A_{standard}$, represents the response area of N-nitrosamines in the working standard.

Results and Discussion

Temperature Profile of Microwave Cooked Bacon. A rapid increase in temperature of the surface of the empty susceptor was observed during heating in the microwave oven. Temperatures in excess of 100°C (212°F) were observed within 30 seconds. A final temperature of 220°C (430°F) was obtained after microwave heating for 3.5 minutes (Figure 1). Less heat was generated in the transparent package. A consistent elevation in temperature was found during heating in this package, however, no temperature greater than 110°C (230°F) was observed after 6 minutes of microwaving (Figure 1). The susceptor package converted microwave energy into heat more efficiently.

Lower temperatures were observed when the susceptor packages were loaded with bacon during cooking. After 2 minutes of cooking, the temperature recorded was 113°C (235°F) (Figure 2). Higher temperatures were obtained as the cooking times increased, 166°C and 211°C (331°F and 412°F) at 2.5 and 3 minutes of microwave cooking, respectively.

During microwaving, temperatures observed on the surface of the transparent package material containing bacon were greater than those observed for the empty package alone. Higher temperatures were associated with longer cooking times (Figure 3). Temperatures of 140°C to 159°C (284° to 319°F) were achieved during heating of the transparent packages containing bacon for 4 minutes in a microwave oven.

A susceptor acts as a secondary source of heat during microwave heating (3), because it absorbs microwave energy and expresses it as heat to the food product. However, transparent materials do not respond to the microwave. Bacon heated on susceptors reached the same "degree of doneness" in a shorter cooking time compared to bacon cooked in transparent packages.

Formation of N-nitrosamines in Cooked Bacon. N-nitrosamine concentrations in bacon processed in the MSU Meat Laboratory and cooked under three different conditions are presented in Table I. NPYR was not detected in bacon in the transparent packaging material until it

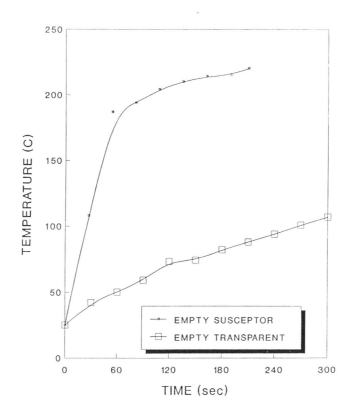

Figure 1. Temperature profile on the surface of the empty susceptor and transparent package heated in a 700W microwave oven.

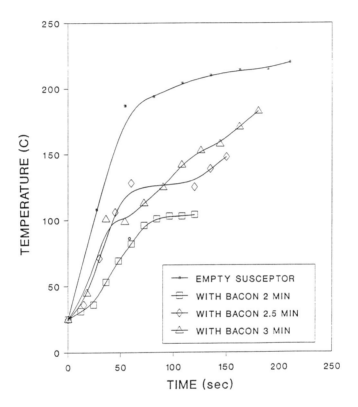

Figure 2. Temperature profile on the surface of susceptors empty and containing bacon during heating in a 700W microwave oven.

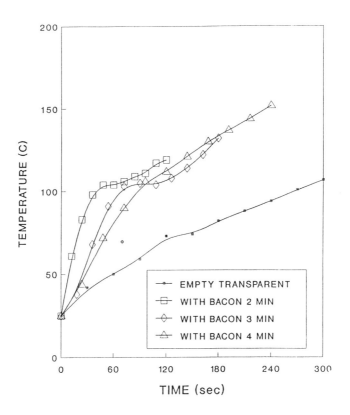

Figure 3. Temperature profile on the surface of transparent packages empty and containing bacon during heating in a 700W microwave oven.

was cooked for 4 minutes. However, with the susceptor material, NPYR was detected after 2.5 minutes of cooking in the microwave. In both procedures, NPYR formation in bacon increased with cooking time. NDMA was also present in the microwave-cooked bacon, but the concentrations detected generally followed no established pattern. This is consistent with literature information regarding the variability of NDMA concentrations in cooked bacon (9). In a similar study with commercial bacon purchased locally, the concentration of NPYR was greater in the fried samples than in those cooked in the microwave (Table II). NPYR was not formed in bacon cooked on the susceptor for 2 minutes and in the transparent package for less than 3 minutes.

In order to compare the effects of cooking procedure on N-nitrosamine formation, it was considered important to compare N-nitrosamine concentration in bacon samples cooked to the same "degree of doneness". This was accomplished visually and it was determined that bacon cooked in susceptor packages for 2.5 minutes and in transparent packages for 3 minutes had the same "degree of doneness" as fried bacon (Figure 4). Cooking for longer periods in the microwave produced overcooked samples. A comparison of the N-nitrosamine data in Tables I and II indicates that bacon fried in a skillet contained more NPYR than the bacon cooked by either microwave method to the same doneness. More NPYR was formed by susceptor cooking than by cooking in the transparent packages in all samples based on the same degree of doneness. It is reasonable to assume that the greater quantities of NPYR formed in susceptor cooking is due to the higher cooking temperatures involved. It is widely believed that NPYR in cooked bacon is formed through the initial nitrosation of proline, followed by thermal decarboxylation of N-nitrosoproline to NPYR (11,24). Similarly, the higher concentrations of NDYR in fried bacon relative to the microwave-cooked samples also is illustrative of the time-temperature dependent mechanism of the formation of this N-nitrosamine in bacon.

The effect of cooking procedure on concentration of NDMA in the bacon samples was not as evident as for NPYR. This is consistent with previous observations (9,20) and reflects the variability in NDMA concentrations in bacon.

Effect of Nitrite Concentrations on N-Nitrosamine Formation in Bacon. Formation of NPYR in bacon is not only influenced by cooking temperature and time (10,11,12,18), but also by the ingoing concentration of sodium nitrite during processing (9). To further demonstrate the influence of cooking method on NPYR formation, bacon was processed with an ingoing sodium nitrite concentration of 200 mg/kg meat to facilitate NPYR formation (Table III). Frying produced the greatest concentration of NPYR followed by susceptor package and transparent package microwave cooking when all samples were cooked to the same degree of doneness. Greater concentrations of NPYR were produced in

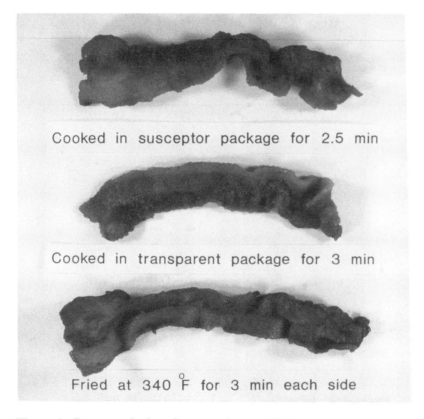

Figure 4. Bacon cooked to the same degree of "doneness" by frying and microwave cooking in susceptor and transparent packages.

Table I.   N-Nitrosamine Concentrations in Bacon[a,b] Processed in the
            MSU Meat Laboratory and Cooked Using Three Different
            Methods

Cooking Method	NDMA (μg/kg)	NPYR (μg/kg)
FRYING 3 min/ side	0.7 ± 0.1[c]	4.6 ± 0.4
TRANSPARENT 2 min	ND[d]	ND
3 min	1.5 ± 0.2	ND
4 min	0.7 ± 0.0	3.0 ± 0.1
SUSCEPTOR 2 min	ND	ND
2.5 min	1.4 ± 0.1	2.3 ± 0.1
3 min	1.1 ± 0.0	5.6 ± 0.5

[a]   The average residual sodium nitrite concentration was 21.4 ± 4.6
       mg/kg meat. Ingoing target level was 120 mg/kg meat.
[b]   Recoveries of the internal N-nitrosamine standard (NAZET) ranged
       from 92 to 117 ± 11%.
[c]   N-nitrosamine concentrations were the average of duplicate
       determinations of three replicated experiments.
[d]   None detected.

Table II.   N-Nitrosamine Concentrations in Commercial Bacon[a,b]
Cooked by Three Different Methods

Cooking Method	NDMA (μg/kg)	NPYR (μg/kg)
FRYING 3 min/ side	0.5 ± 0.2[c]	6.8 ± 1.3
TRANSPARENT		
2 min	0.2 ± 0.1	ND[d]
3 min	1.5 ± 0.2	ND
4 min	2.1 ± 0.2	1.3 ± 1.0
SUSCEPTOR		
2 min	0.5 ± 0.1	ND
2.5 min	0.3 ± 0.0	1.9 ± 0.1
3 min	0.6 ± 0.0	4.0 ± 0.8

[a]   The average residual sodium nitrite concentration was 37.2 ± 16.4 mg/kg meat.

[b]   Recoveries of internal N-nitrosamine standard (NAZET) ranged from 93 to $104 \pm 11\%$.

[c]   N-nitrosamine concentrations were the average of duplicate determinations of three replicated experiments.

[d]   ND = none detected.

Table III.   N-Nitrosamine Concentrations in Bacon Processed with Two
Different Ingoing Levels of Sodium Nitrite and Cooked by
Three Different Methods to the Same Degree of Doneness[a,b]

Cooking Method	NDMA (μg/kg)	NPYR (μg/kg)
120 mg/kg nitrite		
FRYING 3 min/ side	1.5 ± 0.1^c	4.6 ± 0.1
SUSCEPTOR 2.5 min	0.4 ± 0.1	1.1 ± 0.1
TRANSPARENT 3 min	0.3 ± 0.0	1.4 ± 0.4
200 mg/kg nitrite		
FRYING 3 min/ side	2.2 ± 0.0	7.4 ± 0.7
SUSCEPTOR 2.5 min	0.6 ± 0.1	1.3 ± 0.6
TRANSPARENT 3 min	0.3 ± 0.0	0.9 ± 0.1

[a]   Residual sodium nitrite levels in bacon processed with 120 and 200
mg/kg were 47.8 ± 2.7 and 63.1 ± 1.9 mg/kg, respectively.

[b]   Recoveries of N-nitrosamine internal standard (NARET) ranged
from 96 to 109 ± 10%.

[c]   N-nitrosamine concentrations were the average of duplicate
determinations of three replicated experiments.

fried bacon samples processed with 200 mg/kg sodium nitrite compared to bacon samples processed with 120 mg/kg sodium nitrite. However, no significant difference in NPYR concentrations was observed for microwave cooking of the bacon samples processed with the two different nitrite levels. NDMA concentrations in fried bacon also appeared to increase with processing nitrite level, although concentrations were not influenced by the microwave cooking procedures.

Conclusions

This study has demonstrated that higher temperatures are attained during the cooking of bacon in susceptor packages compared to cooking in transparent packages. This higher temperature was reflected in greater concentrations of NPYR when the bacon samples were cooked to the same degree of doneness. The following general conclusions can be drawn from this study:

1. Microwave cooking reduces the formation of NPYR in cooked bacon when prepared to the same degree of doneness as in frying.

2. NPYR concentrations in microwave-cooked bacon increased with cooking time regardless of which packaging technique was used.

3. NPYR concentrations in fried bacon were greater in the bacon samples processed with 200 mg/kg sodium nitrite. No significant differences in NPYR concentration were obtained when the respective bacon samples were cooked by microwave. This is due to the low levels of NPYR produced during the two cooking procedures.

4. Greater concentrations of NPYR were detected in commercial bacon than in the bacon produced in the MSU Meat Laboratory. This may be the result of only using three commercial samples in the study, or possibly because the MSU bacon was processed from fresh pork bellies of known history. It is well established that NPYR formation in bacon is directly related to preprocessing procedures (9).

Literature Cited

1. Anonymous *Packaging Week* 1988, 4, p. 16.
2. Martin, J.R. *Paper, Film and Foil Converter* 1988, 62, p. 55.
3. Perry, M.R. In *Food Product-Package Compatibility*; Gray, J.I., Harte, B.R. and Miltz, J., Eds.; Technomics Publishing Co: Lancaster, PA, 1987, pp. 178-199.
4. Dilberakis, S. *Food and Drug Packaging* 1987, 51, p. 18.
5. Rosenkranz, T.; Higgins, P. *Microwave World* 1987, 8, p. 10.
6. Densford, L. *Food and Drug Packaging* 1988, 52, p. 3.

7. Labuza, T.P. *Cereal Food World* 1988, 33, p. 523.
8. Rice, J. *Food Processing*, 1988, 49, p. 60.
9. Skrypec, D.J.; Gray, J.I.; Mandagere, A.K.; Booren, A.M. *Food Technol.* 1985, 39, p. 74.
10. Pensabene, J.W.; Fiddler, W.; Gates, R.A.; Fagan, J.C.; Wasserman, A.E. *J. Food Sci.* 1974, 39, p. 314.
11. Bharucha, K.R.; Cross, C.K.; Rubin, L.J. *J. Agric. Food Chem.* 1979, 27, p. 63.
12. Wasserman, A.E.; Pensabene, J.W.; Piotrowskov, E.G. *J. Food Sci.* 1978, 43, p. 276.
13. Fiddler, W.; Pensabene, J.W.; Fagan, J.C.; Thorne, E.J.; Piotrowski, E.G.; Wassermann A.E. *J. Food Sci.* 1974, 39, p. 1070.
14. Sen, N.P.; Donaldson, B.A.; Seaman, S.; Iyengar, J.R. and Miles, W.F. *J. Agric. Food Chem.* 1976, 22, p. 397.
15. Fiddler, W.; Pensabene, J.W.; Piotrowski, E.G.; Phillips, J.G.; Keating, J.; Mergens, W.; Newmark, H.L. *J. Agric. Food Chem.* 1978, 26, p. 653.
16. Pensabene, J.W.; Fiddler, W.; Miller, A.J.; Phillips, J.G. *J. Agric. Food Chem.* 1980, 28, p. 966.
17. Sen, N.P.; Iyengar, J.R.; Donaldson, B.A.; Panalaks, T.; Mandagere, A.K.; Booren, A.M.; Pensabene, J.W. *J. Agric. Food Chem.* 1974, 22, p. 540.
18. Theiler, R.F.; Sato, K.; Aspelund, T.G.; Miller, A.F. *J. Food Sci.* 1981, 46, p. 691.
19. Theiler, R.F.; Sato, K.; Asperlund, T.G.; Miller A.F. *J. Food Sci.* 1984, 49, p. 341.
20. Reddy, S.K.; Gray, J.I.; Price, J.F.; Wilkens, W.F. *J. Food Sci.* 1982, 47, p. 1598.
21. A.O.A.C. *Official Methods of Analysis*; Association of Official Analytical Chemists, Washington, DC, 1984, Vol. 14; pp. 437-439.
22. Magee, P.N.; Montesano, T.; Preussmann, R. In: *Chemical Carcinogens*; American Chemical Society, New York, NY, 1976, 491-625.
23. Gray, J.I.; Bussey, D.M.; Dawson, L.E.; Price, J.F.; Stevenson, K.E.; Owens, J.L.; Robach, M.C. *J. Food Sci.* 1981, 46, p. 1817.
24. Gray, J.I.; Collins, M.E.; MacDonald, B. *J. Food Protect.* 1978, 41, p. 31.
25. Lee, M.L.; Gray, J.I.; Pearson, A.M.; Kakuda, Y. *J. Food Sci.* 1983, 48, p. 820.

RECEIVED May 8, 1991

Chapter 12

Thermodynamics of Permeation of Flavors in Polymers

Prediction of Solubility Coefficients

G. Strandburg[1], P. T. DeLassus[2], and B. A. Howell[3]

[1]Center for Applications in Polymer Science, and [3]Department
of Chemistry, Central Michigan University, Mt. Pleasant, MI 48859
[2]Barrier Resins and Fabrication Laboratory, The Dow Chemical Company,
1603 Building, Midland, MI 48674

The use of polymers as food packaging materials has experienced rapid and continuous growth in recent years. This trend is likely to endure (1). Polymer packages have evolved from simple food wraps to sophisticated containers which have additional demands placed upon them. The package must participate in the flavor management of the food.

Flavor degeneration is a complex issue, and various modes of loss can occur. Of primary importance is oxidation resulting from invasion of oxygen from the atmosphere through the container and subsequent reaction with the food. Elimination of this type of flavor degeneration requires that the package restrict the transport of oxygen. It must also limit the flux of water vapor. It is critical that moist foods do not dehydrate and that dry foods remain dry.

Flavor degeneration can also result from migration of large molecules, typically organic solvents and flavor and aroma molecules. These will contain four to twelve carbon atoms and may contain functional groups as well. Contamination of the food from an external source is one mode of flavor alteration. An example of this process is migration of perfumes used in detergents and health and beauty aids into food packages during storage. Ingress of solvents used in printing inks and adhesives are also representative of this method of flavor degeneration. An area of greater concern is the migration of flavor and aroma molecules from the food. Flavor and aroma molecules are present in extremely small quantities; often the total concentration is less than one part per million. However these molecules are responsible for the unique flavor of a particular food, and small losses often result in dramatic off taste. Losses can be from broad removal of all flavors or an imbalance caused by selective removal of only a few flavor components.

Flavor losses that result from the interaction with the polymer package can be classified into two categories. First, losses occurring by permeation or migration through the package, and, second,those from sorption or scalping by the container. Many plastic packages contain a barrier polymer to minimize flavor losses. Barrier

0097–6156/91/0473–0133$06.00/0

polymers are polymers that inhibit mass transport. Suppression of mass transport must be non-selective; it must be effective for both large and small molecules. The purpose of this study was to investigate methods to quantify these losses in typical packaging polymers. More important, given the growth of plastics as food packaging materials coupled with the complexity of flavor make-up, it would be advantageous to be able to predict mass transport parameters of flavor molecules in commercial polymers. Included in this work are semi-empirical methods to predict mass transport parameters for a variety of molecules in selected polymers.

The permeation process can be described as a multi-step event. First, collision of the penetrant molecule with the polymer is followed by sorption into the polymer. Next, migration through the polymer matrix by random hops occurs, and finally, desorption of the permeant from the polymer completes the process. The process occurs to eliminate an existing chemical potential difference. Steady-state mass transport in polymers with a constant chemical potential follows Fick's first law of diffusion. For a polymer having a thin film geometry, Fick's first law can be written as,

$$\Delta M_x = \frac{P\ A\ \Delta p_x}{\Delta t\ \ L} \qquad (1)$$

where $\Delta M_x / \Delta t$ is the transport rate of material x through a film of area A, having a thickness of L, and under a chemical potential created by pressure difference across the film of Δp_x. P is the permeability coefficient and is a steady-state parameter. In this study S.I. units will be used, thus the permeability coefficient will be reported as $kg*m/m^2*s*Pa$. Equation 2 is used to describe the mass transport process.

$$P = D * S \qquad (2)$$

The permeability as discussed earlier is a steady-state parameter. It consists of two component parts, the diffusion coefficient (D) and the solubility coefficient (S).

The diffusion coefficient is a kinetic parameter. It is a measure of how fast transport events will occur. It reflects the ease with which a penetrant molecule moves within a polymer host. More specifically, the time required to reach steady-state transport is provided by this parameter. It provides an estimate of the effective depth of penetration of a permeant into the matrix as a function of time. Knowledge of the diffusion coefficient is crucially important in applications where steady state is not reached.

The solubility coefficient is a thermodynamic parameter. It is a measure of the concentration of penetrant molecules that will be in position to migrate through the polymer. The solubility coefficient is an equilibrium partition coefficient for distribution of the penetrant between polymer and vapor phase such that the following equation holds.

$$C_x = S * p_x \qquad (3)$$

Where C_x is the concentration in the polymer and p_x is the partial pressure of x in the vapor phase. S is analogous to a reciprocal Henry's law constant. The solubility coefficient is dependent on many variables. Of primary importance is the condensability of the penetrant. The solubility coefficient is also a function of the thermodynamic interaction of the penetrant with the polymer. Losses of flavor/aroma molecules into the polymer package can be quantified from a knowledge of the diffusion and solubility coefficients. While numerous models exist to estimate the diffusion process in polymers, little pertaining to the thermodynamic parameter has been reported. This study was undertaken to introduce fundamental understanding of the thermodynamics involved in the mass transport process. Arguments throughout will be supported with experimental data.

EXPERIMENTAL

A variety of methods exist for measuring permeability, diffusion, and solubility coefficients for large molecules. These methods have been described elsewhere (2). The derivative method was used exclusively in this study. The polymers were studied as mono-layer structures having thin film geometry. Figure 1 represents detector response as a function of time for this technique. The detector response in the derivative method reflects the mass transport rate $\Delta M_x/\Delta t$. At t_0 a film, free of penetrant molecules, is exposed to penetrant vapor on one side, hereafter the upstream side. The downstream side is continually swept by a carrier gas to a detector. As a result the downstream side of the film is always at zero partial pressure of penetrant. Initially the mass transport rate is below the detector limits. As time progresses the transport rate increases to allow detection. The transport rate rises steadily through a transient region and eventually levels off at a steady-state rate at which time $\Delta M_x/\Delta t$ is constant. After calibration, the permeability coefficient can be obtained from the detector response using equation 1. The diffusion coefficient was determined from equation 4 (3).

$$ D = \frac{L^2}{7.2 * t_{1/2}} \tag{4} $$

In this expression D is the diffusion coefficient, L is the film thickness and $t_{1/2}$ is the time required to reach one half the steady-state mass transport rate. Once the permeability coefficient and the diffusion coefficient have been obtained, the solubility coefficient can be calculated using equation 2.

Figure 2 is a schematic of the instrument used for this study. It has been described in detail elsewhere (4). The instrument consists of a gas handling system integrated with a mass spectrometer. The gas handling system contains the plumbing, glassware, mixing pumps, permeation cell, and switching valves necessary for a permeation experiment. The entire system is housed in an insulated gas chromatograph air bath oven. Temperature control is from sub-ambient to $150°C \pm 1°C$.

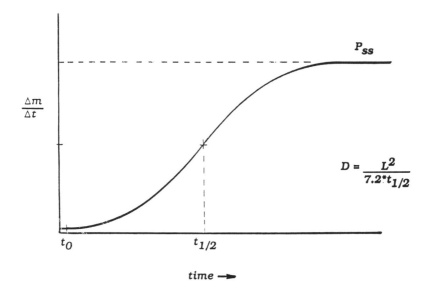

Figure 1. Typical Detector Response Curve for the Derivative Technique for the Determination of Mass Transport Parameters.

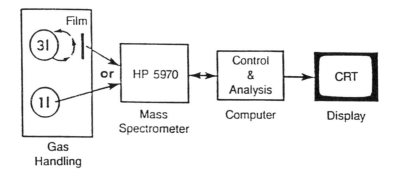

Figure 2. Schematic Diagram for the Permeation Instrument.

The mass spectrometer employed is a Hewlett Packard Model 5970B Mass Selective Detector. The mass spectrometer can be operated in scan mode or in a selected-ion monitoring mode. Data acquisition, storage, and editing are accomplished with a Hewlett Packard 59970C Chem Station. The mass spectrometer was used in two ways. First, mass spectra of permeants and solvents were generated. Ion fragments from the spectra were then selected for the permeation experiments. The criterion for selection was based on the most populous ion fragments that were not degenerate with those from the dilution solvent used in the calibration standard. After selection of the ions, the mass spectrometer is programmed for the selected ion monitoring (SIM) mode. In SIM the mass spectrometer monitors only ions that have been selected. Typically three ion fragments were monitored in each experiment. Subsequent to collecting permeation data all flow paths and flasks were checked to be free of penetrant. This was accomplished by sequentially setting the switching valves to each position required in an experiment while monitoring the visual display. If substantial amounts of permeant were detected, the experiment was aborted and the gas handling system was purged with nitrogen carrier gas until free of any contaminants. After the gas handling system was deemed to be free of permeant, the experiment was started.

The first portion of the experiment was for calibrating the mass spectrometer. Initially, a baseline was established for the calibration process. After the baseline was obtained, a known quantity, typically 2-3 μl, of the calibration standard (0.5% vol/vol dilution) was added to the one-liter flask. This quantity generated a partial pressure of penetrant of about 0.3 Pa (three parts-per-million mole/mole) in nitrogen varying somewhat with the formula weight and density of the penetrant. Following a sufficient mixing time, the calibration standard was sent to the mass spectrometer. After the detector signal for the calibration standard leveled off, a known quantity, usually 2-3 μl, of neat permeant was added to the three-liter flask and was allowed to mix. The partial pressure in the three-liter flask was about 20 Pa (200 parts-per-million mole/mole) of permeant in nitrogen. After a constant calibration response was obtained, the standard was exhausted and a baseline for the experiment was established. At a recorded time, t_0, the penetrant vapor was allowed to circulate on one side of the film while the opposite side was continually swept to the mass spectrometer. The experimental progression was followed by visual observation of the monitor.

Following the experiment the gas handling system was purged with carrier gas to remove excess permeant. Data from the experiment were retrieved, and conversion to a hard copy for analysis completed the experimental process.

Figure 3 is a total ion chromatogram for the transport of ethyl valerate through a vinylidene chloride copolymer film at 105°C. The initial curve represents the calibration portion of the experiment and is followed by the permeation portion.

The experimental design for this study had three parts. First, a rigorous set of experiments to measure the mass transport coefficients for a complete homologous family of compounds through one polymer film was conducted. Many of these experiments were done at small temperature intervals, and multiple experiments were performed at selected temperatures. The data base was used to identify and quantify tendencies as the penetrant was changed in a controlled manner while the polymer was held constant. The goal of this experimentation was to establish the ability to predict the mass transport characteristics for other compounds within this family. The permeants used in this study were linear esters from Flavor and Fragrances Kit No. 1 from The Aldrich Chemical Company. These compounds were used without further purification. The esters used appear in Table I along with selected physical properties.

Table I. Linear Esters for Permeation Studies

Ester name	Boiling Point (°C)	Purity (%)	CAS Number
Methyl butyrate	102	98+	623-42-7
Ethyl propionate	99	97+	105-37-3
Ethyl butyrate	120	98+	105-54-4
Propyl butyrate	142	98+	105-66-8
Ethyl valerate	144	98+	539-82-2
Ethyl hexanoate	168	98+	123-66-0
Ethyl heptanoate	188	98+	106-30-9
Ethyl octanoate	206	98+	106-32-1
Hexyl butyrate	208	98+	2639-63-6

The polymer chosen for the first part of the experimental design was a copolymer of vinylidene chloride and vinyl chloride henceforth co-VDC. It is a commercial polymer used as a barrier layer in food packaging applications. The co-VDC has extremely small diffusion coefficients and thus low steady-state mass transport rates. Elevated temperatures were required (85°C - 105°C) for more timely results.

The second part of the experimental process was to obtain mass transport coefficients for the ester family in another polymer. In this portion of the process a subset of the esters was used. The esters included methyl butyrate, propyl butyrate, ethyl hexanoate, and ethyl heptanoate. The polymer used in this section was low density polyethylene (LDPE). LDPE is not a barrier for the transport of these compounds but represents an important packaging material. Another reason for including LDPE was that mass transport data already exist for selected molecules through this material. Hence, mass transport measurements of properly chosen esters through LDPE allow for a check of the experimental technique.

The third part of this study was an attempt to expand the general observations of the mass transport coefficients for the esters to analogous processes for other homologous families. The compounds chosen to complete this section of the study were linear ketones and n-alkanes. These compounds were chosen since the size and shapes are similar to those of the linear esters. Therefore differences observed for these compounds would be expected to arise from functional group differences. The ketones and alkanes were obtained from The Aldrich Chemical Company and used without further purification. Selected members from these families and some characteristic parameters appear in Table II. Comparing results from the three families should provide insight with respect to mass transport for many different polymer/penetrant systems.

Table II. Ketones and n-Alkanes Used in Permeation Studies

Compound Name	Boiling Point (°C)	Purity (%)	CAS Number
2-Butanone	80	99+	78-93-3
2-Hexanone	127	99+	591-38-8
3-Octanone	167	99	106-68-3
n-Heptane	99	99	142-82-5
n-Octane	126	99+	111-65-9
n-Decane	174	99+	124-18-5

RESULTS AND DISCUSSION

The solubility coefficient is a vapor/polymer equilibrium partition coefficient. The extent of distribution of penetrant vapor in the polymer is dependent, to a large extent, on the energy required to keep the penetrant in the vapor phase. Within a homologous series of compounds an increase in methylene units along the backbone will also increase the heat of vaporization and lead to larger solubility coefficients. From Equation 3 it can be seen that larger solubility coefficients reflect more favorable partitioning into the polymer of the respective penetrant molecules. Figure 4 presents the solubility coefficients as a function of boiling points (K) of the linear ester series in co-VDC at 85 °C. In this plot boiling points represent, as a first approximation, heats of vaporization using Trouton's rule. While it is advantageous to use heats of vaporization to explain the thermodynamic process, these have generally not been reliably reported. Boiling points, on the other hand, are known with reasonable certainty. Figure 5 is a plot of solubility coefficients versus boiling point of the penetrant molecule of the ester series in low density polyethylene at 30 °C. Again the trend is for the solubility coefficient to increase as the number of methylene units in the ester backbone increases.

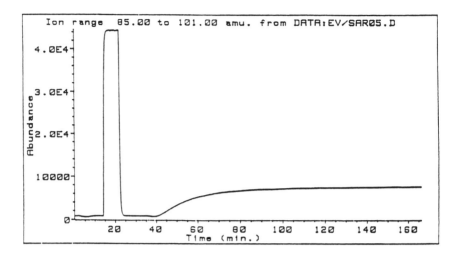

Figure 3. Total Ion Chromatogram for the Permeation of Ethyl Valerate
through a Vinylidene Chloride Copolymer Film at 105°C.

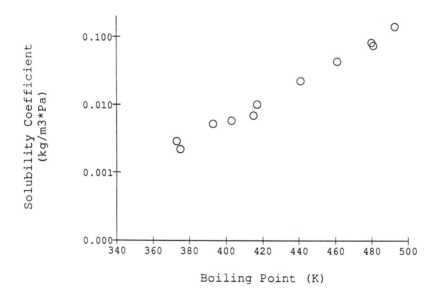

Figure 4. Solubility Coefficients for the Linear Esters in co-VDC at 85°C.

Figure 6 is a plot of the solubility coefficients versus boiling points for three homologous families, esters, n-alkanes, and ketones in co-VDC at 85°C. As before, the trend is consistent within each series but much lower solubility coefficients are realized for the n-alkane family at a specified boiling point (heat of vaporization). For example n-decane, 3-octanone, and ethyl hexanoate boil at about the same temperature (442 K) and, thus, have about the same heats of vaporization; yet, they have much different solubility coefficients in the co-VDC polymer. This observation can be explained by examination of the van't Hoff type equation:,

$$S_T = S_0 \exp(-\Delta H_S/RT) \qquad (5)$$

where S_0 is a pre-expontial factor, R is the ideal gas law constant, T is the absolute experimental temperature, and ΔH_S is the heat of solution. The thermodynamics involved in the heat of solution process can be expressed as;

$$\Delta H_S = \Delta H_{cond} + \Delta H_{mix} \qquad (6)$$

or

$$\Delta H_S = -\Delta H_{vap} + \Delta H_{mix} \qquad (7)$$

The solubility coefficients are not only dependent on heats of vaporization but the partitioning is also dependent on the thermodynamics of the resulting mixture. If an estimation of ΔH_S and S_0 can be obtained, then predictions of S_T may be possible using Equation 5. The heats of vaporization may have been determined or may be estimated from the boiling points. Heats of mixing are generally not available. However estimations of ΔH_{mix} can be obtained using Hildebrand's approximation (5). The heat of mixing in this model is considered either to be zero or positive and can be expressed as;

$$\Delta H_{mix} = k|\delta_1 - \delta_2|^2 \qquad (8)$$

where δ_1 and δ_2 are the solubility parameters of the penetrant and polymer respectively and k is the product of the volume fraction of each component. For this study a value of k equal to one has been most useful for correlating data. The solubility parameter is the square root of the cohesive energy density for low molecular weight species. It is the sum of the intermolecular energies per unit volume. Solubility parameters for many polymers and low molecular weight compounds are available. If the solubility parameters are not available, they can be calculated using equation 9,

$$\delta = ((\Delta H_{vap} - RT)\rho/F.W.)^{1/2} \qquad (9)$$

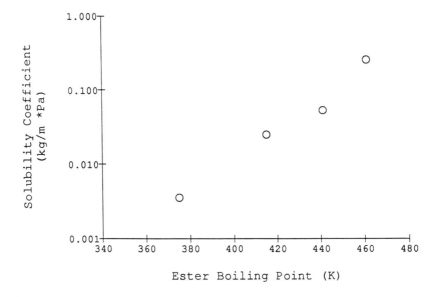

Figure 5. Solubility Coefficients for the Linear Esters in LDPE at 30°C.

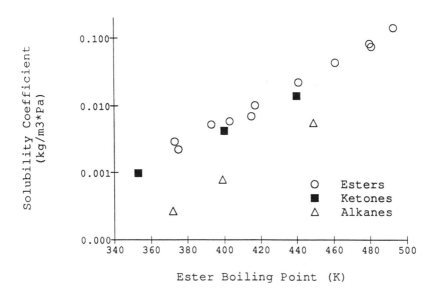

Figure 6. Solubility Coefficients for the Linear Esters, n-Alkanes, and Linear Ketones in co-VDC at 85°C.

where ρ is the density of the liquid with formula weight, F.W. This relationship is valid if the vapor behaves approximately as an ideal gas (6). The solubility parameter has units of (energy/volume)$^{1/2}$. For this study the solubility parameters are listed in units of (cal/cc)$^{1/2}$.

Table III lists the experimental data for the solubility coefficients for selected penetrants in co-VDC at 85 °C along with the heat of solution calculated from Equation 7 for each penetrant. Figure 7 shows graphically the results in this table. If a value of 10.3 (cal/cc)$^{1/2}$ is used for the solubility parameter for the co-VDC all the data points fall on a single linear plot of solubility coefficients versus heat of solution when a heat of mixing term is included.

Table III. Heats of Vaporization, Heats of Mixing, Calculated Heats of Solution, and Solubility Coefficients at 85 °C for Selected Penetrants in co-VDC Film

Penetrant Molecule	δ_1 (cal/cc)$^{1/2}$	$-\Delta H_{vap}{}^1$ Calculated	$\Delta H_{mix}{}^1$	$-\Delta H_S{}^1$	S_{85} kg/m^{3}Pa
Methyl butyrate	8.9	8.98	0.22	8.76	$2.2 \times 10^-$
Ethyl propionate	8.9	8.93	0.22	8.71	2.9×10^{-3}
Ethyl butyrate	8.6	9.43	0.38	9.05	5.2×10^{-3}
Ethyl valerate	8.5	9.96	0.48	9.48	1.0×10^{-2}
Propyl butyrate	8.5	10.0	0.48	9.52	9.1×10^{-3}
Ethyl hexanoate	8.2	10.6	0.73	9.87	2.2×10^{-2}
Ethyl heptanoate	8.1	11.0	0.88	10.1	4.6×10^{-2}
Ethyl octanoate	8.0	11.5	1.2	10.3	8.7×10^{-2}
Hexyl butyrate	8.0	11.5	1.2	10.3	7.4×10^{-2}
$\underline{n}$-Heptane	7.5	8.93	1.15	7.78	2.7×10^{-4}
$\underline{n}$-Octane	7.6	9.58	1.18	8.40	7.8×10^{-4}
$\underline{n}$-Decane	7.8	10.7	1.22	9.48	5.6×10^{-3}
2-Butanone	9.3	8.47	0.09	8.38	9.7×10^{-4}
2-Hexanone	8.5	9.60	0.40	9.20	4.2×10^{-3}
3-Octanone	8.2	10.6	0.70	9.90	1.4×10^{-2}

^{1}(kcal/mole)

The value used for the solubility parameter agrees favorably with values determined experimentally elsewhere (7). The utility of this method is that, if solubility parameters can be obtained or calculated from existing thermodynamic data for low molecular weight compounds, then solubility coefficients for these compounds in co-VDC can be estimated. Table IV lists the calculated heats of solution for d-limonene and dipropyl disulfide in co-VDC, the estimated S_{85}, and the experimentally determined S_{85}. The estimated values agree favorably with the experimental values.

Table IV. Estimated Solubility Coefficients from Calculated Heats of Solution for d-Limonene and Dipropyl Disulfide in a Vinylidene Chloride Copolymer at 85 °C

Penetrant Molecule	$\delta 1$ $(cal/cc)^{1/2}$	Calculated $-\Delta H_S^1$	Estimated S_{85} $(kg/m^3 Pa)$	Experimental S_{85} $(kg/m^3 Pa)$
d-Limonene	8.0	9.4	8.0 x 10-3	6.1 x 10^{-3}
Dipropyl-disulfide	8.5	10.2	4.7 x 10^{-2}	4.5 x 10^{-2}

[1](kcal/mole)

This method includes a polymer dependent term, and it should to hold for estimating the solubility coefficients for many polymer/penetrant combinations at constant temperature if S_0 is approximately a constant. Table V lists the calculated heats of solution for selected penetrants in a hydrolyzed copolymer of ethylene and vinyl acetate henceforth EVOH (using a value of $11.7(cal/cc)^{1/2}$ for the solubility parameter of EVOH). Experiments with EVOH were done at relative humidities of less than 10% since mass transport is dramatically affected by increasing humidities. Also included in Table V are the experimental values for S_{85} as well as the estimated values obtained from extrapolation from Figure 7. Again agreement between the predicted and experimental values is quite good.

Table V. Estimated Solubility Coefficients from Calculated Heats of Solution for Selected Penetrants in EVOH at 85 °C

Penetrant Molecule	δ_1 $(cal/cc)^{1/2}$	Calculated $-\Delta H_S^1$	Estimated S_{85} $(kg/m^3 Pa)$	Experimental S_{85} $(kg/m^3 Pa)$
Methyl butyrate	8.9	8.10	5.0 x 10^{-4}	4.0 x 10^{-4}
Propyl butyrate	8.5	8.47	1.1 x 10^{-3}	9.3 x 10^{-4}
Ethyl hexanoate	8.2	8.60	1.5 x 10^{-3}	1.1 x 10^{-3}
Ethyl heptanoate	8.1	8.64	1.6 x 10^{-3}	2.6 x 10^{-3}
2-Butanone	9.3	7.85	3.6 x 10^{-4}	1.7 x 10^{-4}
2-Hexanone	8.5	8.4	8.0 x 10^{-4}	1.0 x 10^{-3}
3-Octanone	8.2	8.70	1.8 x 10^{-3}	1.4 x 10^{-3}

[1](kcal/mole)

Because of the uncertainty in extrapolations of solubility coefficients to lower temperatures, it is not possible to compare results obtained experimentally for co-VDC with those for LDPE obtained at much lower temperatures. However this method can be used to estimate solubility coefficients in polypropylene from solubility coefficients determined in LDPE. Table VI is a collection of calculated heats of solution and S_{30} values experimentally determined for selected penetrants in LDPE. Figure 8 is a graphical representation of the data presented in Table VI. The accepted literature value (8) for the solubility parameter, and the one used in the calculations, for low density polyethylene is $8.0(cal/cc)^{1/2}$. The quality of the data for LDPE is not as good as that for co-VDC, but a good linear plot of solubility coefficients versus calculated heats of solution is again obtained. In these models the solubility parameter can be used as an adjustable parameter to give improved fits if desired. This was not attempted here.

Table VI. Calculated Heats of Solution and Experimentally Determined Solubility Coefficients for Selected Penetrants in Low Density Polyethylene at 30°C

Penetrant Molecule	Calculated $-\Delta H_S$ (kcal/mole)	Experimental S_{30} (kg/m^{3}Pa)
Methyl butyrate	9.0	3.4×10^{-3}
Propyl butyrate	10.0	2.3×10^{-2}
Ethyl hexanoate	10.6	5.8×10^{-2}
Ethyl heptanoate	11.0	2.2×10^{-1}
2-Butanone	8.3	9.2×10^{-4}
2-Hexanone	9.6	7.4×10^{-3}
3-Octanone	10.6	8.5×10^{-2}
<u>n</u>-Heptane	8.9	2.8×10^{-3}
<u>n</u>-Octane	9.6	1.5×10^{-2}
<u>n</u>-Decane	10.7	9.0×10^{-2}

Table VII lists calculated heats of solution and the experimental S_{30} values for ethyl butyrate, ethyl hexanoate, and d-limonene in polypropylene (using 8.3 $(cal/cc)^{1/2}$ as the solubility parameter for polypropylene) (8). Also appearing in Table VII are the predicted S_{30} values extrapolated from Figure 8. There is good agreement between the predicted and experimentally determined values.

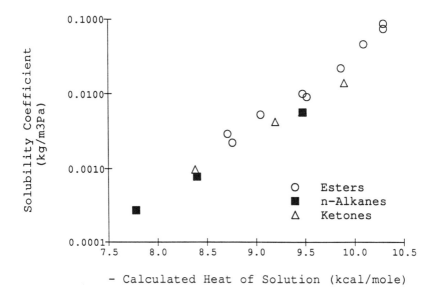

Figure 7. Solubility Coefficients as a Function of Calculated Heats of Solution for the Esters, Alkanes, and Ketones in co-VDC at 85 °C.

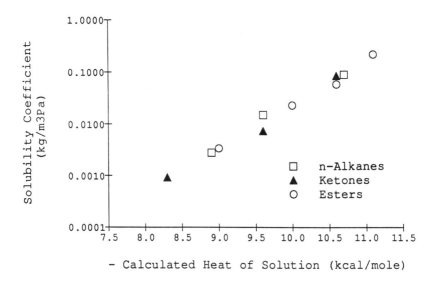

Figure 8. Solubility Coefficients as a Function of Calculated Heats of Solution for the Esters, Alkanes, and Ketones in LDPE at 30 °C.

Table VII. Calculated and Predicted Thermodynamic Data for Selected Penetrants in Polypropylene at 30°C

Penetrant Molecule	Calculated $-\Delta H_S$ [1]	Estimated S_{30} (kg/m^3Pa)	Experimental S_{30} (kg/m^3Pa)
Ethyl Butyrate	9.43	1.1×10^{-2}	1.8×10^{-2}
Ethyl Hexanoate	10.6	7.7×10^{-2}	2.0×10^{-1}
d-Limonene	10.4	5.6×10^{-2}	3.0×10^{-1}

[1] (kcal/mole)

In the examples above the degree of crystallinity was assumed to be constant for co-VDC and the hydrolyzed ethylene vinyl acetate copolymer, and for LDPE and polypropylene. It has previously been shown that the crystalline regions of a polymer above T_g are excluded volumes (9,10). The solubility coefficient can be corrected for crystallinity using equation 10:

$$S_X = S_A(1-\alpha) \qquad 10$$

where S_X is the effective solubility coefficient of the crystalline polymer, S_A is the solubility coefficient of the amorphous polymer, and α is the volume percent that is crystalline.

The ability to predict the thermodynamic coefficient along with existing models for prediction of the diffusion coefficients allows estimations of losses of critical flavor/aroma components for a variety of packaging materials. This study provides a fundamental understanding to the sorption and diffusion processes. As new food packaging materials enter the market the results of this study will be invaluable for predicting the magnitude of flavor/aroma losses to be expected

LITERATURE CITED

1. Hotchkiss, J. H., In "Food Packaging Interactions", Hotchkiss, J. H., Ed.,; ACS Symposium Series No. 365; American Chemical Society: Washington, D. C., 1988, pp. 1-10.

2. Felder, R. M.; Huvard, G. S. In "Polymers - Part C: Physical Properties;" Fava, R. A., Ed.; Methods in Experimental Physics; Academic Press: New York, 1980, Vol. 16: pp. 315-377.

3. Pasternak, R. A.; Schimscheimer, J. F.; Heller, J. J., J. Polym. Sci. Part A-2 1970, 8, p. 467.

4. Tou, J. C.; Rulf, D. C.; DeLassus, P. T., Anal. Chem. 1990, 62, p. 592.

5.   Hildebrand, J.M.; Prausnitz, J. M.; Scott, R. L. "Regular and Related
      Solutions, van Nostrand-Reinhold: New York, 1970.

6.   Rudin, A. "The Elements of Polymer Science and Engineering"
      Academic Press, London, 1982.

7.   Wesling, R. A. "Polyvinylidene Chloride" Gordon and Breach,
      N.Y., 1977.

8.   Burrell, H. In "Polymer Handbook;" Brandrup. J.; Immergut, E. H., Eds.;
      Wiley and Sons: New York, 1975. pp. IV 337 - IV 359.

9.   Michaels, A. S.; Bixler, H. J., J. Polym. Sci. 1961, L, p. 393.

10.   Michaels, A. S.; Parker, R. B., J. Polym. Sci. 1959, XLI, p. 53.

RECEIVED June 25, 1991

Chapter 13

Determination of Flavor–Polymer Interactions by Vacuum-Microgravimetric Method

Ann M. Roland and Joseph H. Hotchkiss

Institute of Food Science, Stocking Hall, Cornell University, Ithaca, NY 14853

A method to quantify the sorption of aroma compounds by food-contact polymers was developed. The method consisted of a microgravimetric balance which was contained in a vacuum chamber. Polymer samples were placed on the balance in the sealed chamber and out gassed to 10^{-4} torr. Small amounts of aroma compounds (d-limonene or d,l-linalool) were admitted to the evacuated chamber and the mass increase in the polymer resulting from compound sorption as well as the pressure within the vessel recorded by computer. Additional aroma compound was admitted to the chamber after equilibrium was achieved and a new equilibrium mass-pressure combination recorded. Sorption isotherms were constructed for each compound-polymer combination from individual equilibria. Solubility coefficients as well as predictions of the amount of aroma compound sorbed at a given aroma concentration (i.e., partial pressure) were made. Results indicated that this technique may be a useful predictor of individual flavor-polymer interactions at flavor concentrations and amounts which are found in foods. A hypothetical prediction of limonene sorption by a cereal box liner was made using data generated by this technique.

Sorption of food aromas by packaging materials is factor in quality alteration during storage (1). Changes in both intensity and character of flavor mixtures resulting from exposure to polymers can be detected by the human nose (2). These changes are due to absorption of certain organic compounds that make up complex aromas by food-contact polymers. This phenomena has been termed "scalping."

No standard method for predicting the amount of sorption occurring for any polymer-flavor combination has been developed. Several sorption studies have utilized sorption cell methods where

0097–6156/91/0473–0149$06.00/0

the penetrant is dissolved or suspended in a liquid and brought into
contact with polymer. Essential oils, pure compounds (2), or complex
mixtures such as orange juice are used (3, 4). The polymer and
liquid are either placed in an inert container (2, 3), or the liquid
is placed in a polymeric container (4). Sorption is measured by
quantitative evaluation of the aroma compound(s) remaining after
equilibrium or by quantitative headspace evaluation.

In direct weighing methods, specimens are exposed to saturated
vapor, withdrawn, weighed, and returned at intervals. Alternately,
samples immersed in a liquid penetrant are removed, blotted, weighed
and returned. For example, the diffusion of organic vapors into
polystyrene and other polymers has been studied using this method
(5-8). The liquid penetrant method was recently used by Aithal et
al. (9) to study aromatic penetrants in polyurethane membranes.

Short comings of current methods include: the presence of an
air barrier, disturbances during the weighing, and heating of the
specimen by condensation of solvent vapor when weighing is done in
environments at temperatures lower than the liquid penetrant (6).
Measurements at high or saturated vapor pressures may not reflect
sorption from foods where aromas are well below saturation. Such
high penetrant concentrations can result in plastization of the
polymer which changes its sorptive properties.

Recent constant vapor concentration methods employ a vapor
generating/dilution system. A constant concentration of penetrant
vapor stream is produced by bubbling N_2 through liquid penetrant and
diluting with untreated N_2, and passing it over the polymer (10).
The polymer is continuously weighed.

Mohney et al. (11) recorded the sorption of d-limonene vapor
in high density polyethylene/sealant laminant film utilizing this
method. Limonene vapor concentrations of 0.3-7.0 ppm (wt/v) were
used. For the laminant and a d-limonene vapor concentration of 1.5
ppm (wt/v), a solubility coefficient (S) of 7.6 mg/(gxppm) was
found.

Some methods involve relatively large amounts of aroma
components because of the limited sensitivity of detection methods.
High amounts act as infinite reservoirs as compared to amounts of
aroma compounds present in foods. Other system components such as
aqueous media, or solvents used to disperse aroma compounds may
affect the partitioning. There is an added consideration of
disturbances caused by the vapor stream decreasing the potential
sensitivity of mass determinations.

Our objective was to determine specific flavor-polymer
interactions at penetrant levels found in the headspace of foods
(often ppb concentrations) using finite amounts of aroma. We
required a method that was quantitative and predictive.

Materials and Methods

Sorption Apparatus. A Cahn 2000 Electrobalance, control unit,
weighing unit and vacuum chamber were used (Cahn Instruments,
Cerritos, CA; Figure 1). Joints were sealed with o-rings and
Apiezon L high vacuum grease (Apiezon Products Limited, London,
England). The electrobalance stand was mounted on four SM-1 Stabl-

Levl pneumatic mounts (Barry Controls, Burbank, CA). The balance chamber was isolated from the pumping system by a 30 cm LN_2 trap. The chamber was closed off from the pump and trap by a valve after pump down.

Internal pressure was continuously measured (type 600 Barocel Pressure Sensor, Datametrics/Dresser Industries, Inc., Wilmington, MA; 10 Vdc, 10.000 mmHg full range). Ultra-Torr vacuum fittings (Cajon Co. Macedonia, OH) and Teflon high vacuum valves were used for all connections. Sample temperature was continuously recorded (Type J thermocouple). Mechanical (Sargent-Welch Skokie, IL model 1402F) and oil diffusion pumps were used. The vacuum pump oil, Plasma Oil 80 (CVC Products, Inc. Rochester, NY), is rated at a vapor pressure of 8 x 10^{-6} mmHg at 25°C. The diffusion pump oil, Convoil 20 (CVC Products, Inc. Rochester, NY), is rated at a vapor pressure of 10^{-6} to 10^{-8} mmHg at 25°C.

Mass was recorded in millivolts (mV) with a full scale reading of 9995 μg and a sensitivity of 1 μg.

A thermostated 23 $\pm$ 3.5°C insulated room housed the balance. A thermostated water bath (VWR Model 1145, VWR Scientific, San Francisco, CA) circulated temperature controlled water via tubing around the vacuum bottle and sample tube which was held in a water filled Dewar flask. A constant sample temperature of 23.5 $\pm$ 0.5 °C was maintained.

A metal encased ionizing unit (Staticmaster; Nuclear Products Co. El Monte, CA) controlled static electricity inside vacuum chamber.

Polymer sample. Polyethylene (Scientific Polymer Products Ontario, New York, catalog #560, density 0.92 g/cm^3, softening point 107°C) was hot pressed between Mylar sheets (103-105°C; Loomis Engineering Co., Caldwell, NJ; 10,000 lb/in^2). The sample was cooled at room temperature. Samples (Table I) were pierced with a 0.01 mm diameter hangdown wire.

Table I: Polyethylene film samples

Aroma compound	Thickness	Area(cm^2)
d-Limonene	1.8 mil (0.046 mm)	10.0
d,l-Linalool	2.0 mil (0.051 mm)	15.0

Aroma Compounds. (+)-Limonene (97%) (1-methyl-4-isopropenyl-1-cyclohexene) was obtained from Aldrich Chemical Company (Milwaukee, WI). d,l-Linalool (95-97%) (3,7-dimethyl-1,6 octadien-3-ol) was obtained from Sigma Chemical Company (St. Louis, MO).

Sorption experiments were carried out as follows: Liquid penetrant was placed in the penetrant vessel (Figure 1), frozen in LN_2 and outgassed through the vacuum pumping system at 4×10^{-5} mmHg. The penetrant valve was closed. The film sample was suspended from one of the arms of the balance. The thermocouple was positioned within 1 cm of the sample. The mass was determined and the balance mechanically and electronically tared. The chamber was evacuated (4 $x10^{-5}$ mmHg) until a constant polymer mass was obtained. The valve isolating the main chamber from the pumping system was then closed. The penetrant vessel valve was then briefly opened and closed to admit a small finite amount of aroma into the glass vacuum vessel. Once the equilibrium mass was attained a second dose of vapor was admitted. A new equilibrium sorption mass was then measured at the higher aroma vapor concentration. This procedure was repeated at increasing vapor concentrations with a single polymer sample (Figure 2). For determinations at the saturation vapor pressure the penetrant valve was left open.

Temperature, pressure and mass were recorded every 2 sec by a DT2805 data acquisition board, DT707-T screw terminal panel with cold junction (Data Translation, Inc. Marlboro, MA) and a Leading Edge PC computer.

The ideal gas relationship was used to determine penetrant concentration in the vapor and conversions made according to Barton (12).

Results

The mass (M_t) of penetrant sorbed per unit weight of penetrant free polymer at equilibrium was measured as a function of time and pressure. M_t approached a limiting value (M_∞) at higher vapor pressures indicating sample saturation. The ratio M_t/M_∞ was plotted against $t^{1/2}$ to obtain a sorption curve (Figures 3A, 3B; 13).

Sorption curves at saturation vapor pressures (Figure 3A) are similar to curves obtained using constant vapor pressure systems (11). However, when finite amounts of penetrant were admitted, equilibrium mass overshoots followed by loss of mass to equilibrium was observed (Figure 3B).

A sample temperature rise and decrease of 0.1 to 0.3°C accompanied the mass overshoot (Figure 4). The pressure inside the chamber also increased sharply when the penetrant was introduced (Figure 4).

Equilibrium solubility coefficients (S) were calculated from the equilibrium concentrations of penetrant in the polymer and vapor phase according to:

$$S = C_p/C_v \tag{1}$$

where
  C_p = equilibrium solubility in the polymer ($\mu g/mg$)
  C_v = equilibrium concentration in the vapor phase (ppm)

Figure 5 compares the solubility coefficients of linalool and limonene as a function of vapor concentration. At equal vapor concentrations linalool has a higher solubility than limonene.

The equilibrium vapor concentration may also be expressed as a ratio of the penetrant's partial pressure to its saturation vapor pressure. When this is plotted against mass gained, a sorption

DIAGRAM OF VACUUM MICROBALANCE APPARATUS

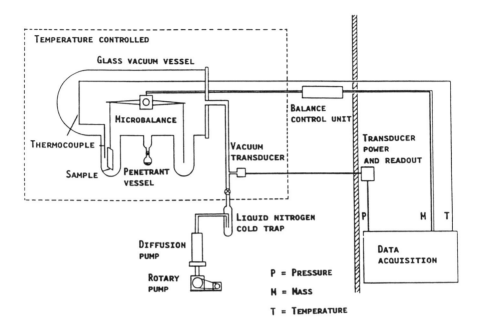

Figure 1:   Sorption Apparatus.

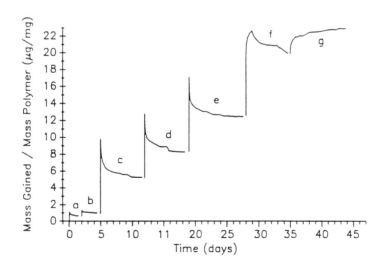

Figure 2:   Mass gain for d-limonene sorption by polyethylene.
Final vapor pressure (mmHg): a 0.101, b 0.154, c
0.293, d 0.413, e 0.549, f 0.684, g 0.752 .

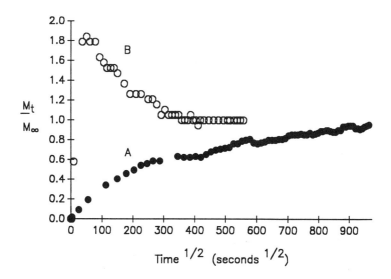

Figure 3:   Sorption of d-limonene by polyethylene at 23.5°C; A
              [limonene]$_{eq}$ = 40.5x10^{-6} mol/L (sat); B [limonene]$_{eq}$
              = 5.5x10^{-6} mol/L.

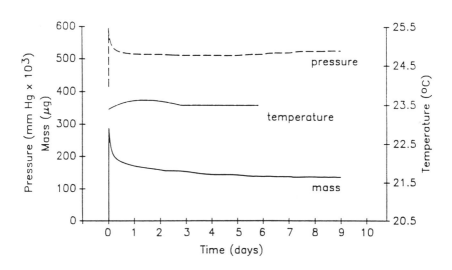

Figure 4:   Mass gain, pressure, and sample temperature for d-
              limonene and polyethylene.

isotherm is constructed for each penetrant/polymer combination (Figure 6).

Discussion

The mass overshoot occurred with each penetrant inlet except at saturation pressures (Figure 2). Similar overshoots in sorption have been observed by Rogers (14) with n-hexane and polyethylene. The cause of such behavior has not been fully explained. It may simply be that opening the penetrant valve allows for a large temporary increase in effective penetrant pressure in the chamber with a resulting overshoot in mass gain. It is also possible that the penetrant condenses on the sample then equilibrates with the vapor phase to reach an equilibrium mass. However, Rogers (14) and Feughelman (15) suggest that a transformation of the polymer may occur during sorption. A large number of weak interchain bonds could be broken by the initial inrush of penetrant. Some of these bonds could be reformed later causing the exclusion of previously sorbed penetrant.

An alternative hypothesis suggests that under conditions where diffusion into thin specimens of high penetrant affinity is rapid, the heat of sorption may cause a large increase in local temperature. The heat of condensation could also cause an increase in sample temperature (16). Using a system of alkanes and polycarbonate Chen and Edin (17) demonstrated that dissolution (sorption) is exothermic and that the heat of solution is linearly related to the heat of condensation. The temperature increase from heat of sorption may locally affect the polymer by thermal expansion or melting of crystalline regions. As the temperature returns to ambient, crystalline or semi-ordered regions may reform and a portion of the penetrant excluded from specimen (14).

The higher solubility of linalool than limonene in polyethylene may be caused by in two factors. Linalool is a more linear, less bulky, molecule than limonene which would facilitate its ability to move into the polymer matrix. Also, its higher boiling point (198°C linalool; 175°C limonene) is indicative of its ability to condense and remain within the matrix. Boiling point has been correlated to a higher solubility by previous workers (18-20). Zobel experimentally determined the solubility of limonene but not linalool in polypropylene. However, we calculated the solubility of linalool in polypropylene using the constants derived by Zobel (20) for alcohol vapor in polypropylene at 25°C (log S = -10.584 + (0.0286 x T); where T= boiling point K). The solubility derived for linalool vapor was 770 g/Nm versus Zobel's experimentally determined value of 510 g/Nm for limonene suggesting that linalool has a higher solubility than limonene in polypropylene. Therefore, our experimental results in polyethylene would likely agree with the order of solubilities for Zobel's determinations in polypropylene.

Solubility coefficients derived from the linear portion of sorption isotherms may be useful approximations of the loss of volatiles from a packaged food. Solubility coefficients (S) were determined as follows:

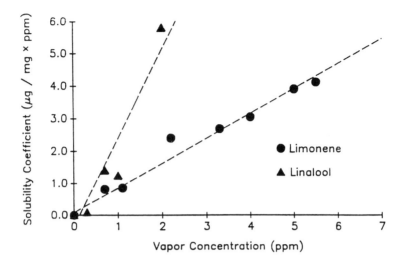

Figure 5:   Solubility coefficients for d-limonene and d,l-
             linalool in polyethylene as a function of vapor
             concentration of penetrant at 23.5°C.

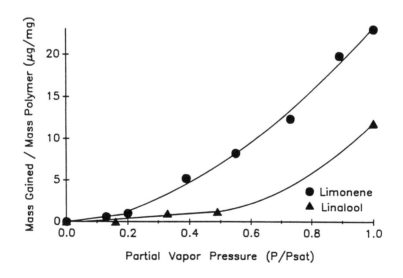

Figure 6:   Sorption isotherms for d-limonene and d,l-linalool
             in polyethylene at 23.5°C.

$$S = C_p/C_v \tag{2}$$

where

C_p = concentration of aroma in the polymer at equilibrium (μg/mg).

C_v = concentration of aroma in the vapor at equilibrium (P/P$_{sat}$).

and

$$C_p \times P = \text{aroma sorbed by package } (\mu g) \tag{3}$$

where

P   = Mass of polymer package (mg)

Solving equation 2 for C_p and substituting into equation 3 and solving for C_p in terms of S:

$$\text{aroma sorbed } (\mu g) = S \times C_v \times P \tag{4}$$

Using this relationship, the equilibrium sorption of limonene from a flavored dry product by a polyethylene cereal liner may be predicted given the following information:

P = 9060 mg PE (48 μm; 45.3 g/m^2; 16 oz. package)
S = 5.0 (μg/mg)/(P/P$_{sat}$) (Figure 6 for limonene)
C_v = 0.1 (10%) (estimated percentage of saturation in headspace)
aroma sorbed (μg) = 5.0 x 0.1 x 9060
= 4530 μg

The mass of aroma sorbed by the package is estimated to be 4530 μg, assuming the aroma concentration in 448 g (16 oz.) of dry product is approximately 30 μg/g (*21*). This represents a loss of 29% of the original 13,400 μg of limonene present.

This example is summarized in Table II along with calculations for limonene and linalool at estimated vapor concentrations in a fruit flavored dry product, (i.e. a breakfast cereal). The analysis assumes a constant aroma vapor pressure in the package and that the partitioning between the cereal and the vapor is very rapid in comparison to the vapor-polymer (i.e., the aroma from the product replaces the sorbed aroma as it is taken up by the package). This maintains a constant concentration of aroma in the vapor at the expense of the aroma content in the product.

Linalool, usually present at 1/10 the concentration of limonene (*21*), may have a greater a potential for affecting aroma due to sorption by a polyethylene package than limonene. The higher solubility of linalool coupled with its higher sensory impact (at least 2 fold that of limonene; *22*) increases the effect of linalool sorption.

Conclusions

Differences in sorption potential combined with differences in concentrations and sensory impact of aroma compounds illustrates that sorption can alter the overall flavor profile (not just intensity) of a packaged product. Solubility coefficients (S) may be

Table II: Example of sorption of the flavor compounds limonene and linalool in a product by a polyethylene package liner

Hypothetical Fruit-Flavored Dry Product

Aroma Compound	Low Conc. Solubility Coefficient $\left(\frac{\mu/mg}{P/P_{sat}}\right)$ S	Vapor Conc. (% of Sat.) C_v	Aroma Sorbed by Package[a] (μg) $C_p \times P$	Aroma Conc. in Product (μg/g)	Aroma Mass in Product[b] (μg)	Aroma Mass Sorbed (%)
d-Limonene	5.0	10 % 1 % 0.5 %	4530 453 226	30	13400	33% 3% 2%
d,l-Linalool	2.4	0.5 % 0.1 %	109 22	3	134	81% 16%

a P = 9060 mg Polyethylene (48 μm; 45.3 g/m²; 16 oz. pkg)
b in 16 oz. (448 g) dry product

useful in predicting the potential of different types of polymeric packaging materials sorb specific compounds. It may be possible to compensate for losses by reformulation. Altering the specific polymers used for a given food may also lead to improved food quality.

The microgravimetric vacuum method provides an accurate means of obtaining sorption isotherms and solubility coefficients for flavor-polymer combinations at penetrant concentrations which are representative of food systems. The apparatus is relatively simple and requires minimal operator attention once the experiments are started. Consecutive determinations may be made on the same sample at different levels of penetrant. The vacuum method minimizes disturbances to the weighing apparatus thus providing greater potential for highest accuracy determination at low penetrant levels where there are smaller mass increases because of sorption.

Literature Cited

1. Farrell, C.J. *Ind.& Eng. Chem. Res.* **1988**, *10(10)*, 1946-1951.
2. Kwapong, O.Y., *Food-package Interactions: The Sorption of Aroma Compounds by Polymeric Materials;* M.S. thesis, Cornell University, Ithaca, NY. 1986.
3. Hirose, K.; Harte, B.R.; Giacin, J.R.; Miltz, J.; Stine, C. In *Food and Packaging Interactions;* Hotchkiss, J.H., Ed.; American Chemical Society: Washington, DC, 1988, Ch. 3; pp 28-41. 1988.
4. Durr, P.; Schobinger, U.; Waldvogel, R. *Alimenta.* **1981**, *20*, 91-93.
5. Crank, J.; Park, G.S. *Trans. of the Faraday Soc.* **1949**, *45*, 240-249.
6. Park, G.S. *Trans. of the Faraday Soc.* **1950**, *46*, 684-697.
7. Park, G.S. *Trans. of the Faraday Soc.* **1951**, *47*, 1007-1013.
8. Laine, R.; Osbourn, J.O. *J. Appl. Polymer Sci.* **1971**, *15*, 327-339.
9. Aithal, U.S.; Aminabhavi, T.M.; Cassidy, P.E. In *Barrier Polymers and Structures;* Koros, W.J., Ed.; American Chemical Society: Washington, DC, 1990, Ch. 19; pp 351-376.
10. Baner, A.L.; Hernandez, R.J.; Jayaraman, K.; Giacin, J.R. *Current Technol. in Flexible Packaging.* **1986**, *ASTM STP 912*, 49-62.
11. Mohney, S.M.; Hernandez, R.J.; Giacin, J.R.; Harte, B.R.; Miltz, J. *J. Food Sci.* **1988**, *53(1)*, 253-257.
12. Barton, A.F.M. *CRC Handbook of Solubility Parameters and Other Cohesion Parameters;* CRC Press, Inc.: Boca Raton, FL, 1983.
13. Fujita, H. 1961. *Adv. Polymer Sci.* **1961**. *3*, 1-47.
14. Rogers, C.E. In *Physics and chemistry of the organic solid state;* Fox, D.; Labes, M.M.; Weissberger, A., Eds.; Interscience Publishers A Div. of John Wiley & Sons, Inc.: New York, NY, 1965, Ch 6, Vol 2.
15. Feughelman, M. *J. Appl. Sci.* **1959**, *2(5)*, 189-191.
16. Crank, J.; Park, G.S. In *Diffusion In Polymers;* Crank, J.; Park, G.S., Eds.; Academic Press: New York, NY, 1968, Ch 1, pp 1-39.

17.   Chen, S.P.; Edin, A.D. *Polymer Eng. Sci.* **1980**, *20(1)*, 40-50.
18.   van Amerongen, G.J. *J. Polymer Sci.* **1950**, *5(3)*, 307-332.
19.   Barrer, R.M.; Barrie, J.A.; Slater, J. *J. Polymer Sci.* **1958**, *27*, 177-197.
20.   Zobel, M.G.R. *Polymer Test.* **1985**, *5(2)*, 153-165.
21.   Furia, T.E.; Bellanca, N. *Fenaroli's Handbook of Flavor Ingredients*; CRC Press, Inc.: Cleveland, OH, 1975, 2nd Ed., Vol. 1.
22.   Buttery, R.G.; Black, D.R.; Guadagni, D.G.; Ling, L.C.; Connolly, G.; Teranishi, R. *J. Agric. Food Chem.* **1974**, *22*, 773-777.

RECEIVED June 20, 1991

Chapter 14

Determination of Food–Packaging Interactions by High-Performance Liquid Chromatography

Amrik L. Khurana[1] and Chi-Tang Ho[2]

[1]Whatman Specialty Products, Inc., 9 Bridewell Place, Clifton, NJ 07014
[2]Department of Food Science, Cook College, New Jersey Agricultural
Experiment Station, Rutgers University—The State University
of New Jersey, New Brunswick, NJ 08903

Interaction of polymeric packing materials such as
polyvinyl acetate (low and heavy carbon load), polymethacrylic acid (heavy carbon load) and polytriphenyl methyl methacrylate (heavy carbon load) with
food components like ascorbic acid, niacin, phenylalanine and caffeine was studied by immobilizing the
polymers on a silica support and using a food based
solvent, like water, as the mobile phase. The mechanism of interactions was determined from enthalpy of
sorption data. The Gibb's free energy and activity
of coefficient data were used to determine the amount
or magnitude and kind (weak or strong) of interactions.

Several kinds of polymeric materials are being used currently to
package foods and drugs. The migration of residual materials from
packaging into food, or of the nutrients from food into the packaging matrix can cause problems. The loss of orange flavor from
orange juice packaged into polymeric plastic bottles is a common
consumer complaint. It is, therefore, very important to study the
interaction of food components with the packaging matrix. Inverse
gas chromatography (1-2) has been used to study the interaction of
volatile food components with packaging materials. Use of HPLC to
investigate interaction of nonvolatile polymers, such as polyvinyl
alcohol, has been reported (3). The thermodynamic parameter, such
as enthalpy of adsorption, was used to conduct such studies. During
the present investigation, use of HPLC has been further extended to
examine the interaction of polymeric materials such as polyvinyl
acetate (low and heavy carbon load), polymethacrylic acid (heavy
carbon load), polytriphenyl methyl methacrylate (heavy carbon load)
with nonvolatile food ingredients such as ascorbic acid, niacin,
phenylalanine and caffeine. Thermodynamic parameters, such as enthalpy of adsorption, Gibb's free energy and activity coefficient
were used to pursue such studies. The polymeric stationary phases
were covalently bonded to silica support by polymerizing the monomers on bonded silica gel. In the previous publication (3), poly-

0097–6156/91/0473–0161$06.00/0
© 1991 American Chemical Society

vinyl alcohol was immobilized by reacting the same with epoxy bonded silica support.

The vinyl bonded silica (A) for synthesizing low carbon load was prepared by reacting 40 g of spherical silica gel (5u) at 70°C with 3 g of vinyl dimethyl chlorosilane ($CH_2=CH-Si(CH_3)_2Cl$) in toluene by using 15 ml of triethyl amine as a catalyst. In the case of vinyl bonded silica (B) used to synthesize heavy carbon load polymer, 40 g of the 5μ spherical silica was reacted at 70°C with 11 g of vinyl dimethyl chlorosilane in toluene by using 20 ml of triethylamine as a condensing agent.

The low carbon load polyvinyl acetate was synthesized by reacting 6 g of vinyl acetate ($CH_2=CH-COOC_2H_5$) with 15 g of vinyl bonded silica (A) in acetonitrile at 80°C using 0.6 g sodium peroxide as a catalyst. The heavy carbon load polymethacrylic acid and polytriphenyl methyl methacrylate were synthesized by reacting 18 g of methacrylic acid ($CH_2=C(CH_3)COOH$) and 2.5 g of methyl methacrylate ($CH_2=C(CH_3)COOC(C_6H_5)_3$) separately with 25 g of vinyl bonded silica (B) in acetonitrile at 80°C using 3 g of sodium peroxide as the catalyst. The final products were extracted with chloroform in a Soxlet apparatus to remove excess of unattached polymers from silica.

The vinyl bonded silica products were treated with trimethyl chlorosilane in toluene at 80°C to cap residual silanols.

There is a relationship between the retention time and the column temperature. The retention time decreases as the temperature is increased and the relationship between these two can be expressed by Equation 1 (4-7).

$$lnK' = \Delta H°/RT - \Delta S/R + ln\psi \qquad (1)$$

Here K' = capacity factor; T = column absolute temperature; $\Delta H°$ = standard enthalpy change on transferring a solute from the stationary phase; ΔS standard entropy change and ψ = the phase ratio.

The standard enthalpy change or $\Delta H°$ can be derived from the slope of the plots of lnK' vs 1/T (van't Hoff plot) (8). Equations 2 and 3 can be used to determine thermodynamic parameters such as G or Gibb's free energy, and γ, the activity coefficient (9-10).

$$\Delta G = \Delta H° - T \Delta S \qquad (2)$$

$$\Delta G = - RT \ln \gamma \qquad (3)$$

It is possible to determine the nature or mechanism of interactions, i.e. hydrogen bonding and hydrophobic, etc. from $\Delta H°$ values. The negative $\Delta H°$ values indicate an exothermic process or retention by hydrogen bond formation through polar groups. The endothermic adsorption process, in the case of the positive $\Delta H°$ values is generally an indication of hydrophobic interactions (1-2).

The magnitude or extent of interactions can be realized from ΔG or Gibb's free energy values and γ, or the activity coefficient can determine the kind (weaker or stronger) of interactions. The higher γ values, i.e. 1 or more than 1, will indicate the weaker interactions, and the lower γ values, such as 0.9 or less than 0.9, determine the existence of stronger surface interactions (9).

Polyvinyl Acetate (Low Carbon Load)

Table I represents the capacity factor, i.e., ln K' values at various temperatures (T) for ascorbic acid, niacin, caffeine and phenylalanine in case of a low carbon load polyvinyl acetate - silica column. The $\Delta H°$ values derived from slopes of the plots of lnK' against 1/T are also shown in Table I. The polymeric stationary phase can be of low or heavy carbon load if the monomer of the same is polymerized on silica gel having single (A) or multiple vinylic chains (B) on a silanol of Si-OH group.

The positive $\Delta H°$ values exhibit an endothermic adsorption process which is an indication of the interaction of all these probes with the polymeric surface due to hydrophobic forces (1-2). The polar groups like -OH, -COOH and $-NH_2$ in the case of ascorbic acid, niacin and phenylalanine, are not interacting with the surface to form hydrogen bonds.

Table I. lnK' at Various Absolute Temperatures
and Enthalpy of Sorption of
Ascorbic Acid, Niacin, Phenylalanine and
Caffeine on Low Carbon Load Polyvinyl
Acetate Column

Components	lnK'	$1/T \times 10^3$	$\Delta H°$ (K Cal/mole)
Ascorbic Acid			+ 3.50
	0.20	2.90	
	0.55	3.00	
	0.76	3.05	
	0.90	3.10	
Niacin			+ 5.26
	0.90	3.00	
	1.23	3.05	
	1.66	3.10	
	1.90	3.20	
Phenylalanine			+ 4.50
	1.23	2.90	
	1.70	3.00	
	1.90	3.05	
	2.20	3.10	
Caffeine			+ 7.60
	1.90	2.90	
	2.53	3.00	
	3.37	3.10	
	3.80	3.15	

Figures 1A and 1B exhibit resolution of ascorbic acid, niacin, phenylalanine and caffeine at 40°C and 70°C. Caffeine shows more change in retention time as compared to the other components. It shows that caffeine has more surface interaction than all other probes under investigation. The ΔG values at various temperatures

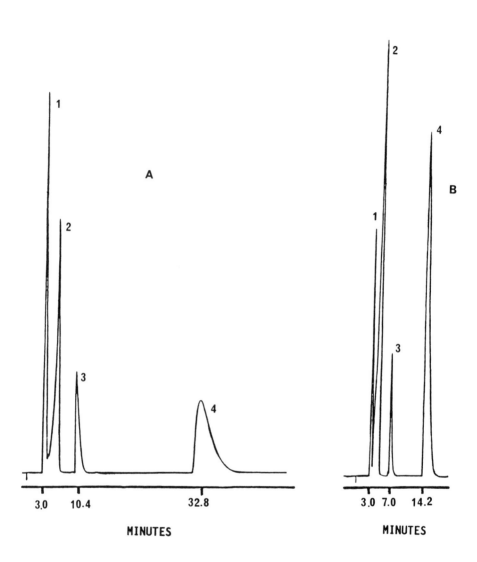

Figure 1
Resolution on Polyvinyl Acetate (Low Carbon Load) PartiSphere
5 Column (25 cm x 4.6 mm) at 40°C (A) and 70°C (B); Mobile
Phase:Water at 0.5 mL/min; λ_{max}: 280 nm;; 1. Ascorbic Acid,
2. Niacin, 3. Phenylalanine, 4. Caffeine.

(Table II) exhibit a similar trend. The order of interaction from ΔG values on low carbon load polyvinyl acetate - silica column can be assigned as follows: caffeine > phenylalanine > niacin > ascorbic acid.

Table II. ΔG Values and the Coefficient of Activity
or γ Values at Various Temperatures in the Case of Ascorbic Acid,
Niacin, Phenylalanine and Caffeine on Low Carbon Load
Polyvinyl Acetate Column

Component	ΔG	γ	T ($^\circ$C)
Ascorbic Acid	0.667	0.410	50
	0.564	0.460	55
	0.364	0.570	60
	0.137	0.800	72
Niacin	1.238	0.150	40
	1.158	0.170	45
	0.891	0.290	55
		0.410	72
Phenylalanine		0.136	50
	1.238	0.150	55
	1.158	0.170	60
	0.891	0.220	72
Caffeine		0.022	45
	2.163	0.034	50
	1.674	0.054	60
	1.303	0.136	72

Table II also shows the activity coefficient values at different temperatures for various components. All these probes have a stronger surface interaction as γ values, in each case, are less than 0.9 (9). These interactions become weaker as the temperature is increased. The stronger surface affinity of these components indicates that their migration from food is a good possibility.

Polyvinyl Acetate (Heavy Carbon Load)

Table III shows the capacity factor or $\ln K'$ values at various absolute temperatures as well as ΔH° values for ascorbic acid, niacin, phenylalanine and caffeine on heavy carbon load vinyl acetate - silica column.
    The enthalpy of sorption shows negative values in the case of ascorbic acid and niacin, and positive values for phenylalanine and caffeine which indicates exothermic and endothermic adsorption processes (1-2). The polar groups such as -OH and -COOH are involved to form hydrogen bonds with the surface in the case of ascorbic acid and niacin.
    Figures 2A and 2B show the resolution of ascorbic acid, niacin, phenylalanine and caffeine at 40°C and 70°C on a heavy carbon load polyvinyl acetate column. A higher retention time is observed in the case of caffeine which indicates more surface interaction for this probe at various temperatures than the rest of the components.

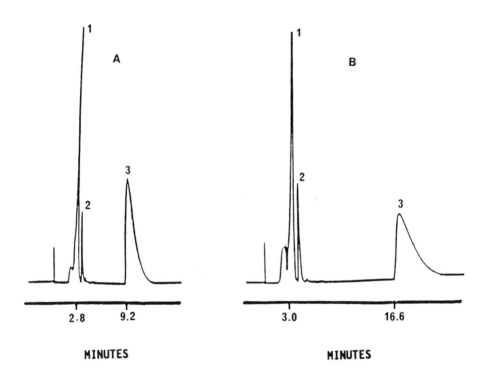

Figure 2
Resolution on Polyvinyl Acetate (Heavy Carbon Load) Parti-
Sphere 5 Column (25 cm x 4.6 mm) at 40°C (A) and 70°C (B);
Mobile Phase: water at 1 mL/min; λ$_{max}$: 280 nm; 1. Ascorbic
Acid and Niacin, 2. Phenylalanine, 3. Caffeine.

Table III.   lnK' at Various Absolute Temperatures and
Enthalpy of Sorption of Ascorbic Acid, Niacin, Phenylalanine
and Caffeine on Heavy Carbon Load Polyvinyl Acetate Column

Components	lnK'	$1/T \times 10^3$	$\Delta H°$ (K Cal/mole)
Ascorbic Acid			- 0.45
	0.73	3.00	
	0.67	3.10	
	0.60	3.30	
	0.55	3.40	
Niacin			-0.25
	0.60	3.10	
	0.55	3.30	
	0.50	3.40	
Phenylalanine			+ 3.30
	0.80	3.20	
	1.10	3.30	
	1.40	3.40	
Caffeine			+ 0.25
	0.73	3.00	
	0.80	3.10	
	0.92	3.20	

Table IV. ΔG Values and the Coefficient of Activity or
γ Values at Various Temperatures in the Case of Ascorbic Acid,
Niacin, Phenylalanine and Caffeine on Heavy Carbon Load
Polyvinyl Acetate Column

Component	ΔG	γ	T (°C)
Ascorbic Acid	0.362	0.570	20
	0.372	0.549	30
	0.392		50
		0.477	60
Niacin	0.321	0.589	30
	0.372	0.566	50
	0.397	0.549	60
Phenylalanine	0.815	0.247	20
	0.662		30
	0.490	0.450	40
	0.362	0.560	50
Caffeine	0.662	0.332	20
	0.572	0.399	40
	0.539	0.432	50
	0.490	0.477	60

ΔG or free energy values at various temperatures (Table IV)
confirms this fact.   The coefficient of the activity or γ values at
different temperatures (Table IV) shows that all these probes have
values less than 0.9 which indicates their stronger surface interactions (9).

Δ G or free energy values at various temperatures (Table IV) confirms this fact. The coefficient of the activity or γ values at different temperatures (Table IV) shows that all these probes have values less than 0.9 which indicates their stronger surface interactions (9).

Polymethacrylic Acid

Tavle V shows the lnK' values at various column absolute temperatures, as well as enthalpy of sorption of the probes on heavy carbon load vinyl methacrylic acid column.
  Enthalpy values are negative in the case of ascorbic acid, niacin and phenylalanine which indicate exothermic adsorption process. The positive value in the case of caffein exhibits an endothermic adsorption process (1-2). In the case of ascorbic acid, niacin and phenylalanine, the polar groups such as C=O, =N- and NH_2 are involved in forming hydrogen bonds with -COOH groups of the surface.

Table V. lnK' at Various Absolute Temperatures and
Enthalpy of Sorption of Ascorbic Acid, Niacin, Phenylalanine and
Caffeine on Polymethacrylic Acid Column

Components	lnK'	$1/T \times 10^3$	$\Delta H°$ (K Cal/mole)
Ascorbic Acid			- 0.175
	0.30	2.90	
	0.30	3.00	
	0.25	3.10	
	0.20	3.30	
Niacin			- 0.127
	0.60	2.90	
	0.60	3.00	
	0.60	3.10	
	0.55	3.20	
Phenylalanine			- 2.400
	1.20	2.90	
	1.20	3.00	
	1.10	3.10	
	1.00	3.40	
Caffeine			+ 1.670
	1.30	2.90	
	1.50	3.00	
	1.70	3.10	

Figures 3A and 3B exhibit resolution of the probe molecules on polymethacrylic acid column at 40°C and 70°C. The caffeine has more surface interaction as compared to other components. A small change in retention time of the other components shows that their interaction does not change much as the temperature is increased whereas, in the case of caffeine it decreases rapidly with the rise of temperature. Δ G or Gibb's free energy values at various temperatures (Table VI) verifies this fact. The higher Δ G values of caffeine show that it has more surface interaction than the rest of the probes.

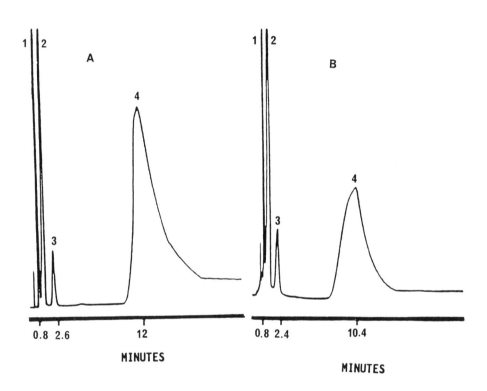

Figure 3
Resolution on Polymethacrylic acid - PartiSphere 5 Column
(25 cm x 4.6 mm) at 40°C (A) and 70°C (B). Mobile Phase:
Water at 1 mL/min; λ_{max} : 280 nm; 1. Ascorbic Acid, 2. Niacin,
3. Phenylalanine, 4. Caffeine.

Table VI also shows the activity coefficient or γ values at various temperatures in the case of ascorbic acid, niacin and phenylaline. The activity coefficient of caffeine changes rapidly with the increase of temperature, whereas it shows a small change in the case of other probes.   The γ values for all these components are less than 0.9, which indicates their stronger surface interactions (9).

Table VI.   Δ G Values and the Coefficient of Activity
or γ Values at Various Temperatures in the Case of Ascorbic Acid,
Niacin, Phenylalanine and Caffeine on
Polymethacrylic Acid Column

Component	Δ G	γ	T (°C)
Ascorbic Acid	0.132	0.803	30
	0.132	0.787	40
	0.161	0.787	50
	0.200	0.770	60
	0.200	0.756	70
Niacin	0.340	0.570	30
	0.340	0.572	40
	0.358	0.572	50
	0.340	0.572	60
	0.358	0.572	70
Phenylalanine	0.582	0.368	20
	0.706	0.333	50
	0.800	0.300	60
	0.800	0.300	70
Caffeine	1.119	0.165	40
	1.091	0.183	50
	0.993	0.224	60
	0.900	0.272	70

Polytriphenyl Methyl Methacrylate

Table VII exhibits lnK' values at various temperatures of the probes and ΔH° derived from slopes of plots of these values.
The negative ΔH° value in the case of ascorbic acid shows an exothermic adsorption process.   The positive value in the case of the rest of the components shows an endothermic process (1-2).   Ascorbic acid is retained through a hydrogen bond formation between -OH and -C=O groups of ascorbic acid and surface, respectively.   The other components, i.e., niacin, phenylalanine and caffeine, are interacting with the surface through hydrophobic forces.
Figures 4A and 4B represent a resolution of the various probes on a polytriphenyl methyl methacrylate column at 40°C and 70°C.   The higher retention time of caffeine at various temperatures indicates that it has more surface interaction than the rest of the components.   The higher ΔG values (Table VIII) confirms this fact.
Table VIII also shows the activity coefficient or γ values at various temperatures of the probes on polyvinyl triphenyl methacrylate column.   All these components have values either 0.9 or less than 0.9 which indicates their stronger surface interactions (9).

Table VII. lnK' at Various Absolute Temperatures and
Enthalpy of Sorption of Ascorbic Acid, Niacin, Phenylalanine
and Caffeine on Polytriphenyl
Methyl Methacrylate Column

Components	lnK'	$1/T \times 10^3$	$\Delta H°$ (K Cal/mole)
Ascorbic ACid			- 0.20
	0.20	2.90	
	0.20	3.00	
	0.20	3.10	
	0.16	3.30	
Niacin			+ 0.40
	0.10	3.00	
	0.10	3.10	
	0.16	3.20	
	0.20	3.30	
Phenylalanine			+ 0.87
	0.70	3.00	
	0.80	3.10	
	0.90	3.20	
	0.96	3.30	
Caffeine			+ 0.18
	0.60	2.90	
	0.75	3.00	
	1.16	3.20	
	1.30	3.30	

Table VIII. ΔG Values and the Coefficient of Activity
or γ Values at Various Temperatures in the Case of Ascorbic Acid,
Niacin, Phenylalanine and Caffeine on Polytriphenyl
Methacrylate Column

Components	ΔG	γ	T (°C)
Ascorbic Acid	0.832	0.870	30
		0.860	40
	0.116	0.835	50
	0.146		60
	0.146	0.800	70
Niacin	0.116	0.819	30
	0.106	0.860	40
	0.076	0.900	50
	0.076	0.923	70
Phenylalanine	0.578	0.382	30
	0.560	0.407	40
	0.489	0.458	60
	0.463	0.540	60
Caffeine	0.801	0.265	30
	0.721	0.313	40
	0.489	0.458	60
	0.415	0.540	70

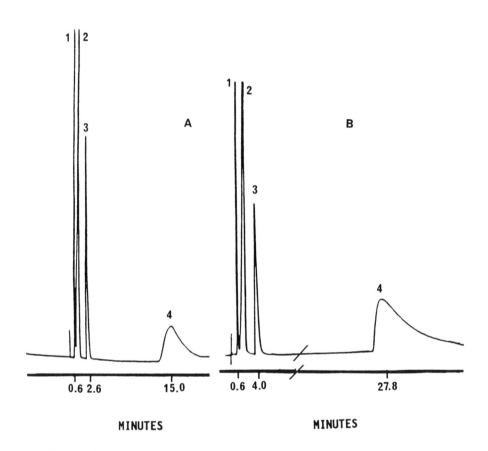

Figure 4
Resolution on Polytriphenyl Methyl Methacrylate - PartiSphere 5
Column (25 cm x 4.6 mm) at 40°C (A) and 70°C (B). Mobile
Phase: Water at 1 mL/min; λ_{max} : 280 nm; 1. Ascorbic Acid,
2. Niacin, 3. Phenylalanine, 4. Caffeine.

The change in interaction with the change of temperature is more pronounced in the case of caffeine and phenylalanine. The activity coefficient values at various temperatures of ascorbic acid on polyvinyl acetate (low and heavy carbon loads), polymethacrylic acid and polytriphenyl methyl methacrylate columns are either equal to or less than 0.9 which indicates stronger surface interaction of ascorbic acid with these polymeric materials. It also predicts the migration of ascorbic acid from food into the polymeric materials. It further shows that all these polymers are unsuitable to be used as containers for packaged food containing ascorbic acid. Similar conclusions can be drawn in the case of niacin, phenylalanine and caffeine.

The present approach of synthesizing and covalently bonding a polymeric stationary phase to a silica support via silylation technique allows us to study and investigate the interaction of any polymer with food and drug components. In the current investigation polymers such as polyvinyl acetate, polymethacrylic acid and triphenyl methyl methacrylate have been investigated, but this technique can be further extended to evaluate commercially available polymeric products used as food and drug packaging materials.

Acknowledgements

New Jersey Agricultural Experiment Station Publication No. D-10205-3-91, supported hy State Funds. We thank Mrs. Joan B. Shumsky for her secretarial aid.

Literature Cited

1. Carrilo, P. J.; Gilbert, S. G.; Daun, H. J. Food Sci. 1988, 53, 1199-1202.
2. Coelho, U.; Miltz, J.; Gilbert, S. G. Macromolecules 1979, 12, 284.
3. Khurana, A. L.; Ho, C-T J. Liquid Chrom. 1989, 12, 1679-86.
4. Hafkenscheid, T. L.; Tomilson, E. J. Chrom. 1976, 122, 47-62.
5. Karger, B. L.; Snyder, L. R.; Eon, C. J. Chrom. 1976, 125, 71-88.
6. Szanto, J. I.; Veress, T. Chromatographia 1985, 20, 596-600.
7. Tijssen, R.; Billet, H. A.; Schoenmakers, P. J. J. Chrom. 1976, 122, 185-203.
8. Halasz, I.; Sebastlan, I. Agnew Chem. Int. Edn. 1969, 8, 453-56.
9. Allen, B.; Chin, H. Pittsburgh Conference, New York, 1990. Abstract Book. The Pittsburgh Conference, 300 Pen Center Boulevard, Suite 332, Pittsburgh, PA 15235, p. 074.
10. Levine, I. N. (Ed.) "Reaction Equilibrium". In Nonideal Systems. Physical Chemistry, 1988, 254, 297-301.

RECEIVED March 20, 1991

Chapter 15

Sensory-Directed Analytical Concentration Techniques for Aroma–Flavor Characterization and Quantitation

Kent Hodges

The Dow Chemical Company, Analytical Sciences, 1897 Building, Midland, MI 48667

"The amount of an odor needed to cause a "nuisance" differs with each person and each separate time the judgement is made. The effects of odors are more psychological than physical."

"Because of the complexity of odors, the odor transmission systems, and odor detection and recognition capabilities, no equipment has been developed to satisfactorily detect and measure odors recognized by the nose."

Quotes such as these occur in a number of olfactory instructional texts. Their content could disincline any scientist who has been involved in trace analytical measurements for many years, and now has the mandate to characterize and quantify aromas and flavors. Most odors and tastes are quantifiable using modern day analytical instrumentation, provided at each step of the analytical separation, concentration and characterization, sensory analysis is employed.

The food packaging industry today is actively involved in measuring odor and taste as it relates to maintaining the integrity of food packaged in various types of containers. Figure 1 identifies some of the interactions that typically could occur between a food, beverage, or pharmaceutical and its container. Trace concentrations of residual solvents, monomers, plasticizers, inhibitors or mold release agents, that are an integral part of the containers manufacturing, could conceivably migrate to the food, thereby imparting an off taste or flavor. In other cases the flavor ingredients of the food can migrate into the container causing the loss of some of its natural aroma profile. Environmental contamination, such as oxygen, water or other volatile odorants could conceivably migrate through the plastic, if appropriate barrier resins are not selected, thereby, adulterating the food and causing it to take on a bad odor or taste. These are the concerns of the food packaging analytical scientist.

Challenges in food packaging today are not necessarily set by the food processor or baker. Todays "olfactory literate" consumer demands convenient, easy to prepare, quality food items with good taste. In addition the containers must be readily disposable, recyclable, convenient to use and aesthetically pleasing. These demands on the food processor or baker challenge a fabricator or

0097–6156/91/0473–0174$06.00/0
© 1991 American Chemical Society

converter. The demand concurrently is accepted by the resin supplier who must offer innovative, high-quality plastics or other materials such that the fabricator can meet these challenges. Todays successful packaging design requires close interaction between the food processor, fabricator, and resin supplier for the following reasons. The food processor understands intimately the chemistries involved in the manufacture and storage of his food. The fabricator is part artist, part engineer, capable of taking resins and molding them into containers which protect and are very aesthetically pleasing. The resin supplier is a plastics chemist, fully understanding the properties of his starting material. Today, these different, complex disciplines must work together to insure the finished product meets the demands of the consumer.

<u>But Why are These Demands so Great Today?</u>
When I first became involved in the measurement of food packaging interactions, 10 years ago, I surveyed the literature to determine what had been written over the past 20 years. In the '60's homo-polymer and co-polymer films and containers were used essentially for short-term, less than one-week, storage under what we might consider low stress conditions, (i.e., refrigeration). During the '70's co-polymers and laminated films became more sophisticated, because the packaging applications demanded longer shelf life (i.e., several weeks). Microwaving introduced a higher stress to the packaging. During the '80's a technology explosion has occurred with multi-layered, barrier containers; some containing 25 different laminated layers to impart characteristics and aesthenics to the package. Shelf life has also been extended, often many months and even years in length. And with the advent of dual-ovenable containers and susceptors, another degree of stress has been imposed. Indeed, very delicate, sensitive food matrices such as cottage cheese, water, orange juice, and even wine are successfully packaged in plastic containers today. Who would have guessed in the '60's that we would be packaging the FDA leachables test media, 30% ethyl alcohol (vodka) in a plastic container? So the demands to produce high quality resins which perform well in the food packaging industry will continue to be extremely challenging.

To meet this challenge at Dow, we have established four emphasis groups to develop a database for food packaging interactions. One group covers the design and engineering as well as materials of construction of such films and containers. Another group concerns itself with the measurement of barrier properties. These are essentially analytical measurements which determine the solubility and diffusivity of migrant molecules through and into plastic films. Product and packaging interactions are covered by another group in which the relationship between aroma and flavor performance and trace analytical measurements are correlated. The fourth group is a food sciences laboratory in which food integrity can be measured, as a function of time for a given packaging application. Common measurements such as moisture, texture and the fate of nutrients that are in a given food product when it is placed in a plastic container, can be determined. A fifth group has recently been added, a sensory testing lab with trained panelists for determining product performance.

Analytical aroma/flavor characterization and quantitation require a knowledge of the three vectors that contributed to "Total Aroma Perception": Concentration, Potency, and Hedonic (Pleasantness or Unpleasantness).

Let's consider trying to measure them or put some quantitative number on "Total Aroma Perception."

The log of odorant concentration is directly proportional to sensory response. As long as we are above the threshold recognition concentration this relationship appears to fit for most odorants. This also points out one of the fundamental challenges we have in the packaging industry. A 100-fold reduction in odorant concentration would only have a 2 unit reduction on the sensory response scale! Therefore, if some package component causes a food product to taste bad, we must reduce its concentration by one hundred fold before its affect would be below threshold recognition concentration.

Figure 2 shows the large range of aroma potency for some organic compounds the human nose can detect. The odor threshold for compounds such as octane and nonane are approximately 1000 ppm. These could be classed as odorants with "weak potency." A compound such as vanilla which is detectable by the human nose at approximately a ppt or 10^9 times lower would be classed as possessing a "strong potency." This variance in odor potency creates one of the analytical separation scientist's dilemmas characterizing odorants in packaging. Solvents used in the manufacture of plastics are selected so as not to contribute taste or odor to the finished package. They have a "weak potency." Sometimes trace levels of these solvents are retained, yet the analytical scheme must separate them from the compounds which are truly impacting the odor or taste problem of the package.

The hedonic tone or pleasantness/unpleasantness impression was for many years thought to be dimensionless. Andrew Drevnicks of the Institute of Olfactory Sciences, has made an attempt to quantify this hedonic tone (Figure 3). Over 250 different types of odors were ranked in terms of pleasantness or unpleasantness by a panel of 350 people. The data was normalized to a scale covering from +4, which is very pleasant, to -4, which is very unpleasant. With some falling in the neutral hedonic rating. We might argue whether the methodology can distinguish between something which is +1 or -1 as agreeable or disagreeable, but I don't think we have any trouble recognizing the difference between the aroma of "fresh baked bread," which has one of the highest universal hedonics, and something which smells like "burnt rubber" or "cadaverous." I think we would all agree packaging films that have a "cadaverous" odor would not sell on the open market.

Concentration techniques for aromas involve a variety of approaches. One technique, headspace gas chromatography, is often described as a viable technique for odor characterization, however, contemporary analytical equipment facilitates measurements only to low ppm. Recalling the sensitivities of human olfaction we described earlier, we need to have detection capabilities on the order of at least a ppb and often a ppt, to truthfully characterize the presence of an odor. Cryofocus, headspace, capillary gas chromatography, or purge and trap techniques, can be applied to measure low ppb by focussing as much as 50-100 milliliters "on-column." Odorant selective concentration techniques such as vacuum steam distillation must be employed to get the 5,000-fold concentration needed to detect these odors. It is very important in all concentration efforts that sensory analysis be performed on the concentrate,

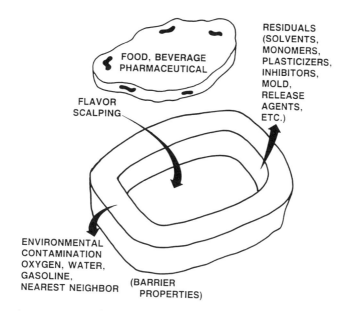

Figure 1. Typical food packaging interactions which can affect odor/taste performance.

Odor Threshold	Compound
10^3	octane, nonane
10^2	ethanol, acetone
10^1	toluene, ethylacetate, methylethylketone
10^0	vinyl acetate, acetic acid
10^{-1}	styrene, mesityloxide, methylmethacrylate
10^{-2}	chlorophenol, eugenol, butanoic acid
10^{-3}	butylacrylate, 2-nonenal, ethylmercaptan
10^{-4}	1-octen-3-one, amylmercaptan, ethylacrylate
10^{-5}	1-nonen-3-one
10^{-6}	vanillin

Figure 2. Example of the wide range of sensory threshold detection concentrations expressed in parts per million (v/v) for some organic compounds.

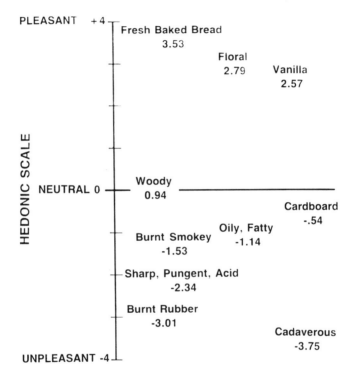

Figure 3. Hedonic Tone (Pleasantness or Unpleasantness of an Odor). (Reproduced with permission from reference 1. Copyright 1984 Air Pollution Control Association.)

thereby proving to the separation scientist that he has indeed isolated the odiforous fraction, without adulteration, before attempting any characterization.

Cryogenic headspace samplers can be used to isolate the odorants from as much as 2 liters of headspace above a given packaging material. The odorants can then be transferred to the front of a capillary gas chromatographic column chilled to liquid nitrogen temperatures. The components are then separated and measured by any one of four or more different types of detectors. Olfaction plays a major role in determining which of the component peaks are sensorially active. Sulfur or nitrogen specific detectors provide a measure of selectivity for certain types of aromas. Matrix isolation, fourier transform infrared spectrophotometers, as well as conventional quadrupole mass spectrometer, are used to provide structural information about the odorant molecule eluting from the column. This is essentially a screening technique, because often we do not obtain the sensitivity to low ppt, which is the concentration range in which we must make measurements.

A modified Nielson-Kryger steam distillation apparatus is shown in Figure 4. A condenser is fixed atop a 2-liter resin pot and cooled to 3° by means of a recirculating bath. Pentane or hexane can be placed in the condenser to extract the volatiles as they are steam distilled from the sample. After the steam distillation is completed, the extract is ready immediately for capillary gas chromatographic analysis.

While steam distillation is most commonly employed to separate and concentrate volatile aromas from food matrixes, packaging films or containers can similarly be extracted. Capillary chromatograms of the extracts from the steam distillation of three packaging films is shown in Figure 5. All of the packaging films are identical except after exhaustive sensory testing, one has an odor intensity which is very weak, designated as a "one". Another has a strong waxy smell, which is ranked as a "three," and another is rated as intermediate, or a "two." In looking at the hydrocarbon response, the chromatograms are quite complex, and any hope of identifying the odorant molecule in this extract would be extremely difficult. Figure 6 shows a simple silica gel class separation that we use to fractionate the extract and essentially separate the hydrocarbon fraction from the more polar compounds present in the sample which cause odor. After depositing the extract on a disposable pipette filled with silica gel, the first fraction is eluted with 2-mL of hexane. This fraction generally consists of saturated hydrocarbons. Next the bed is eluted with 2-mL of 10% methylene chloride in hexane to elute the unsaturates and aromatics. The next fraction is eluted with 2-mL of methylene chloride. This fraction generally contains aldehydes and ketones. The fourth fraction is eluted with 2-mL of methyl-t-butyl ether and generally contains acids, unsaturated ketones, and unsaturated aldehydes. The fifth cut is eluted with 2-mL of methanol which are generally polyfunctional polar type compounds.

After reanalyzing the fractions by capillary gas chromatography, Figure 7, some very interesting observations can be made. The top chromatogram shows the concentrate before class separation, which from a sensory analysis was called "strong, waxy." The first fraction, which ironically looks virtually identical to the concentrate, has no odor! The second fraction has no odor, but the third cut has a very strong waxy odor. The fourth cut has a very slight floral aroma, while the fifth cut, which is the methanol eluant, also has no odor. With this

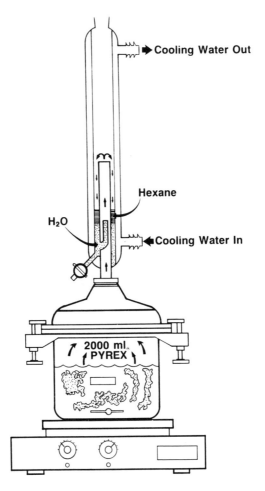

Figure 4. Schematic of a typical simultaneous steam
distillation/hexane extraction apparatus for aroma
concentration.

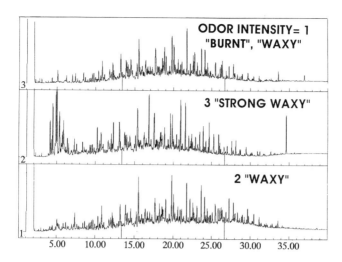

ODOR INTENSITY= 1
"BURNT", "WAXY"

3 "STRONG WAXY"

2 "WAXY"

5.00   10.00   15.00   20.00   25.00   30.00   35.00

Figure 5. Typical capillary gas chromatograms of extracts from the steam distillation of three packaging films.

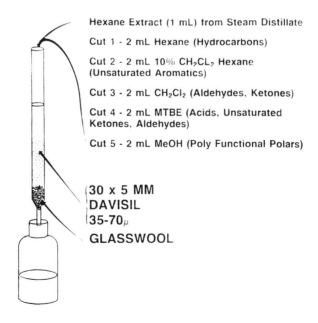

Hexane Extract (1 mL) from Steam Distillate

Cut 1 - 2 mL Hexane (Hydrocarbons)

Cut 2 - 2 mL 10% CH₂CL₂ Hexane
(Unsaturated Aromatics)

Cut 3 - 2 mL CH₂Cl₂ (Aldehydes, Ketones)

Cut 4 - 2 mL MTBE (Acids, Unsaturated
Ketones, Aldehydes)

Cut 5 - 2 mL MeOH (Poly Functional Polars)

30 x 5 MM
DAVISIL
35-70µ

GLASSWOOL

Figure 6. A scheme for the separation of non polar, "low aroma potency" components from the classes of compounds which impact total aroma perception.

simple class separation technique we have two fractions which are readily characterizable to define a "strong waxy" or "slight floral smell." Figure 8 shows what fraction three looks like for each of the packaging films. The chromatograms do indeed have a concentration or intensity profile which correlates very well with their odor intensity. It is easy to see why characterization of these trace level components would be impossible had the hydrocarbon matrix not been removed in the class separation.

Predicting Odor and Taste Performance for Food Packaging Environments
Once the principal odorants in a packaging application are characterized and quantitated, we are ready to do some modeling. Figure 9 shows a model for a typical food packaging application, which would have no odor/taste performance problems. Odorant migration in packaging environments must be below a threshold recognition concentration in a given product to not be detected. If we measure odorant concentration in a food stuff as a function of time, you can see that it increases at a rather rapid rate, the diffusion rate, and finally reaches equilibrium after a period of time, which often takes weeks or months. Figure 10 shows how odorant concentration in a food product may change over time, as a function of temperature. As the temperature under which the product is stored rises, the quantity that might migrate to a food increases dramatically. These curves indicate the difference between warehousing or transportation of packaged food products in summer months, when greater migration may occur compared to the winter months, when the migration would be comparatively much lower. Figure 11 is an example of chocolate chip cookies packaged in a plastic tray and how the concentration of the odorant in the plastic tray plays a key role. The top two curves show that the concentration in the cookies reached 390 ppb and 290 ppb, respectively, under accelerated shelf life studies. The top curve representing the migration at 72°F from a 490 ppm residual tray exceeded the threshold recognition level (TRC) and caused the product to fail sensory analysis 100% of the time. In the second curve when a 350 ppm odorant tray was tested at 72°F, the quantities that migrated were just below the threshold recognition levels, such that the tray failed sensory analysis approximately 50% of the time. However, when a low-residual tray containing 150 ppm was tested, even at 90°F, the quantities which migrated over a 30-week extended shelf life study, were at 135 ppb or at least 2 to 3 times below that which would be considered a threshold recognition level for the odorant in the chocolate cookies. Figure 12 shows a summary for predicting the success or "sensory performance" between a threshold recognition concentration and a measured odorant concentration in certain types of food matrices. Date/oatmeal cookies which have a sensory threshold recognition concentration of 300 ppb of a certain odorant, upon being stored for a given period of time, had measured odorant levels of 33 ppb. Therefore, the sensory performance factor was 10 or 300/33. This indicates a successful packaging application. On the other hand, walnut chocolate-chip cookies, which contain a more lipophilic surface, had a lower threshold recognition concentration, and the quantities that migrated were slightly higher at 70 ppb. This application gave a sensory performance factor of only three which suggested a plastic with lower residuals be used depending on the desired shelf life for this packaging application. If gravy were packaged in this container, the odorant threshold concentration would be 250 ppb and the quantities that migrate would be on the order of 120 ppb which only gives a

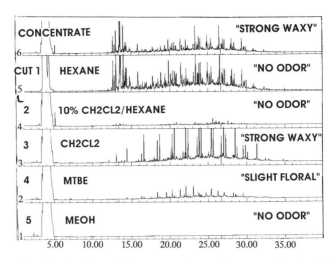

Figure 7. Typical capillary gas chromatograms obtained from a class separation of the hexane extract from the steam distillation of a packaging film.

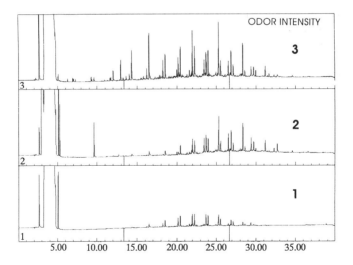

Figure 8. A comparison of the capillary gas chromatograms obtained from fraction three from a silica gel class separation of three packaging films.

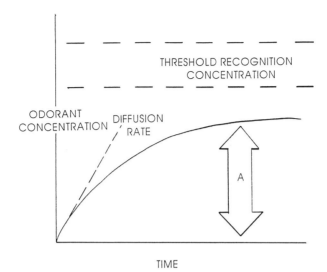

TIME

A - ODORANT CONCENTRATION IN PRODUCT AT EQUILIBRIUM

Figure 9. A model of a typical food packaging application in which the odorant concentration in a food product can change as a function of time.

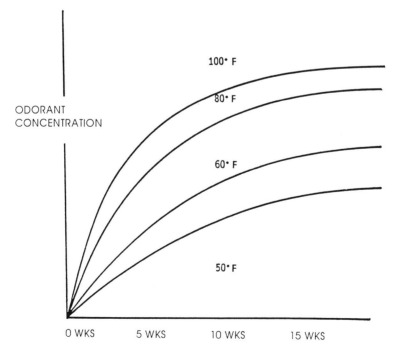

Figure 10. How temperature can affect odorant diffusion rates and subsequent total odorant concentration in a packaged product over time.

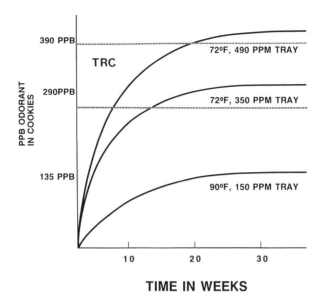

TIME IN WEEKS

Figure 11. Odorant migration to chocolate chip cookies as a function of time and why low odorant residual trays out-perform high residual trays.

Food Matrix	Sensory Threshold Recognition Concentration (ppb)	Measured Concentration (ppb) of Odorant in Food Matrix	Sensory Performance Factor
Date, Oatmeal Cookies	300	33	10
Walnut, Chocolate Chip Cookies	200	70	3
Gravy	250	120	2

Figure 12. Typical sensory performance factors for selected food packaging applications by comparing sensory threshold recognition concentrations to measured odorant concentrations in foods.

performance factor of 2. Therefore, another resin or lower residual resin of the same type would be required for this packaging application.

CONCLUSIONS

Setting quality guidelines and measuring vector relationships of food and plastic package interactions are possible today with sophisticated, sensitive analytical instrumentation. As databases evolve for measuring the parameters affecting "Total Aroma Perception;" concentration of odorant, potency of odorant, and hedonic (pleasantness or unpleasantness) rating of odorant, the ability to model package design will be possible. Cooperative efforts between food processors and bakers, fabricators, and plastics resin manufacturers, will be key to successfully and profitably defining quality guidelines for future food packaging endeavors.

REFERENCES

1. A. Drevnicks, T. Maswrat, and R. Lamm, J. of Air Pollution Control Association (34), July 7, 1984, pp. 752-755.
2. Koszinowski, J. and Piringer, O.; J. Plastic Film and Sheeting, (2), January, 1986, pp. 40-50.

RECEIVED June 6, 1991

Chapter 16

Interaction of Orange Juice and Inner Packaging Material of Aseptic Product

Jinq-Guey Jeng, Ching-Chuang Chen, Shin-Jung Lin, and Chu-Chin Chen

Food Industry Research and Development Institute, P.O. Box 246, Hsinchu, Taiwan, Republic of China

Aseptically processed (brik type, Combibloc) orange juices were stored at 4 C, 25 C and 37 C, for 2 days, 1 week, 2 weeks and 4 weeks. Volatile components absorbed by the inner packaging material (PE) were extracted by ether and analyzed by GC an GC-MS. Non-polar terpene and sesquiterpene compounds (limonene and unknown sesquiterpene compounds) were the major components absorbed by the inner packaging material. Polar volatile compounds such as aldehydes (nonanal and decanal) and alcohols (linalool and alpha-terpineol) were also absorbed by the inner packing material. The amount of volatile components absorbed was significantly affected by the duration of time and temperature during storage. The rate of absorption slowed down after one week of storage. Sensory analysis (4 weeks) indicated that juices stored at 4 C had better sensory score compared to those stored at 25 C and 37 C. Off flavors were generated due to degradation of volatile components at higher storage temperatures.

Orange juice packed in aseptic carton has been commercially viable in Europe for many years. In the Pacific region, juices drinks (ca. 30% juice content) and beverages packaged in aseptic cartons have been popular items in countries such as Japan, South Korea, Taiwan, and Singapore. With the advent of free trade between United States and Pacific countries, consumption of orange juice (100%) in this region is expected to grow. It is also expected that orange juice packed aseptically will become a popular commodity.

0097–6156/91/0473–0187$06.00/0

Although aseptic processing produces high quality products, aseptic orange juice undergoes noticeable flavor changes during storage. These flavor changes may inhibit the market potential of this product (1-4). Marshall et al. (5) indicated orange juice is particularly sensitive to oxidation, non-enzymic browning, and flavor absorption by the polyethylene polymers in contact with the juice. In aseptically packed orange juice, interaction between volatile orange juice components and the inner packaging material were reported by several research groups (6-10).

The present study investigates the volatile components of orange juice absorbed by the inner packaging material, and those volatile components remaining in the juice, after storage at 4 C, 25 C and 37 C, for up to one month.

MATERIALS AND METHODS

Materials. Unless otherwise stated, all other chemicals were of reagent grade. Reagent grade diethyl ether (E. Merck, Darmstadt, W. Germany) was glass distilled. Ethyl cinnamate (internal standard) was obtained from Aldrich (Milwaukee, WI); hydrogen peroxide was obtained from E. Merck (Darmstadt, W. Germany). Orange juice concentrate (63 Brix) was obtained from a local supplier (Chou Chin Industrial Corp., Taiwan). Natural orange flavor was obtained from Fritzsche Dodge & Olcott (No. TW1206, BASF Taiwan Ltd., Taipei, Taiwan). Aseptic packages (ca. 400 ml) were obtained from PKL Corp. (PKL Taiwan Ltd, Taipei, Taiwan). Before being packed with orange juice, each package was first sealed at one end and then sterilized with 30% hydrogen peroxide and dried with hot air under aseptic environment.

Aseptic Processing and Packaging of Orange Juice. Figure 1 shows the scheme of the bench top HTST apparatus and an aseptic packaging system used in the present study. Detailed information about the operation and performance of this system was reported previously (11). Orange juice concentrate was diluted to 11.8 Brix with de-ionized water and the concentration of natural orange juice flavor was adjusted to 0.019% (w/w) (4). The reconstituted orange juice was pasteurized at 85 C for 45 s, cooled down to 15 C and then de-aerated batchwise (ca. 300 ml) before being transferred to aseptic package. The shape of the finished package looks like a tetrahedron instead of the common brik type. Aseptically packed orange juices were divided into three groups and stored immediately at 4 C, 25 C, and 37 C. Sampling dates were on the 0th, 2nd, 7th, 14th, and 31st day, after processing.

Isolation of Volatile Components Absorbed by Inner Packaging Material. Procedures similar to Shimoda et al. (13) were adopted. One top corner of aseptic orange juice package was cut off, juice was transferred to other container for further use. The empty package was rinsed five times with deionized water and air dried. The package was then filled with 250 ml ethyl ether for 24 hr at room temperature while the opened top corner of the package was restricted by a clamp. The ethyl ether extract was concentrated to about 1.0 ml with a vigreaux column. Ethyl cinnamate was added as internal standard.

Isolation of Volatile Components in Aseptic Orange Juice. Procedures similar to Marsili et al. (12) were adopted. Orange juice (100 ml) saturated with NaCl was extracted 3 times with ethyl ether (50 ml), the combined extract was dried over anhydrous sodium sulfate and concentrated to about 1.0 ml with a vigreaux column. Ethyl cinnamate (0.01g/L, 1 ml) was added as internal standard.

GC and GC-MS Analysis. Volatile components extracted from the inner packaging material and orange juice were subjected to gas chromatographic analyses on a Varian 3400 gas chromatograph. A fused silica column with a stationary phase equivalent to Carbowax 20M (DB-WAX+, 30 m x 0.32 mm; J& W Scientific, Folsom, CA) was used. The oven temperature was programmed linearly from 50 to 210 C at 2.0 C/min and was held at 210 C for 40 min. Other operating conditions were as follows: injector and detector temperatures, 250 C; makeup nitrogen flow, 30 mL/min; detector hydrogen flow, 30 mL/min; detector air flow, 300 mL/min. The samples were injected in the split mode with a split ratio of 1/10. The linear velocity of the hydrogen carrier flow was 43 cm/s. Quantitative determinations were made on a PC-based integration system (Chem Lab Corp. Taipei, Taiwan). Linear retention indices were calculated by using n-paraffins (C8-C25; Alltech Associates, Deerfield, IL) as references (14). Capillary gas chromatography-mass spectrometry was also carried out on a Varian 3400 gas chromatograph coupled to a Finnigan Mat mass detector (ITD-800, Finnigan, USA). Analytical conditions of gas chromatograph were the same as above. Analytical conditions of MSD were as follows: ion source temperature, 200 C; ionization voltage, 70 eV; electron multiplier voltage, 2050 V.

Sensory and Statistic Analysis. Sensory evaluations were carried out by 12 experienced panelists. The hedonic 9 point rating scale was adopted. A score of 1 indicated dislike extremely. A score of 4 indicated the border line of acceptance. A score of 9 indicated like extremely. Duplicate juice samples were evaluated each time. Data were analyzed using analysis of variance (PROC ANOVA) and Duncan's multiple range test procedure of the Statistical Analysis System (15).

RESULTS AND DISCUSSION

In order to investigate the interaction between aseptically packed orange juice and the inner packaging material, it is necessary to set up an imitated or real aseptic processing and packaging system for laboratory usages. Conducting such experiment under commercial aseptic processing and packaging equipment is not only expensive but also impractical. It is the reason why FIRDI set up the system shown in Figure 1. Preliminary test results have shown that the quality of aseptic orange juice produced from this system can meet the requirement of commercial processing (11).

Figures 2 and 3 show the gas chromatograms of volatile components of aseptic orange juice absorbed by the inner packaging material, and volatile components remained in orange juice, respectively, after storage at 25 C for a various periods. GC and GC-MS identifications of volatile compounds were accomplished by comparing the retention indices with authentic samples and/or mass spectra with NBS library built in the MS system. Ethyl cinnamate was added as internal standard. Quantitation was based on the relative peak area between sample and internal standard. There were 24 volatile components identified or tentatively identified, which include monoterpenes, sesquiterpenes, terpene alcohols, sesquiterpene alcohols, and aliphatic aldehydes. The identifications were in good accordance with previous reports (4, 12, 16-20).

Tables I, II, and III show the quantitative results of volatile components absorbed by the inner packaging material of aseptic packages stored at 4 C, 25 C and 37 C, for 2, 7, 14 and 31 days. Due to a technical problem, data of volatile components absorbed at 4 C/7th day, was omitted. During the first two weeks, the rate of absorption was fast for juices stored at 25 C and 37 C. The rate of absorption leveled off after two weeks. Compared to juice stored at 25 C and 37 C, only moderate amounts of volatile components were absorbed for juice stored at 4 C.

Tables IV, V, and VI show the quantitative results of volatile components remaining in aseptic orange juice after the aseptic packages were stored at 4 C, 25 C and 37 C, for 0, 2, 7, 14 and 31 days. Within the first 2 to 7 days, the amount of most volatile components remaining in the juice dropped quickly, the rate was faster for juices stored at 25 C and 37 C. Probably due to the oxidation reactions mentioned in previous reports (1-2, 4-6). There was significant reduction of limonene at higher temperature and longer storage time, on the contrary, alpha-terpineol, one of the most important off

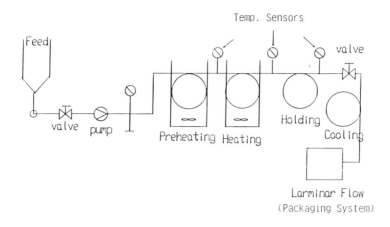

Figure 1. Laboratory aseptic processing and packaging system set up at FIRDI.

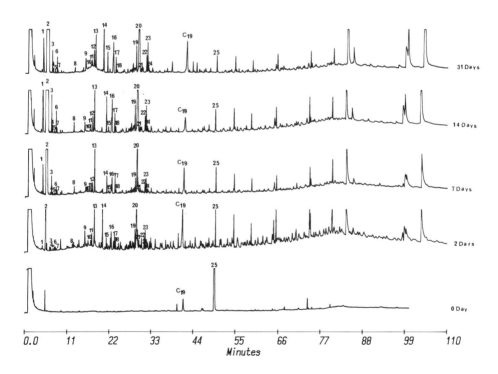

Figure 2. Gas chromatogram of volatile components of aseptic orange juice absorbed by the inner package when stored at 25 C for 0, 2, 7, 14 and 31 days.

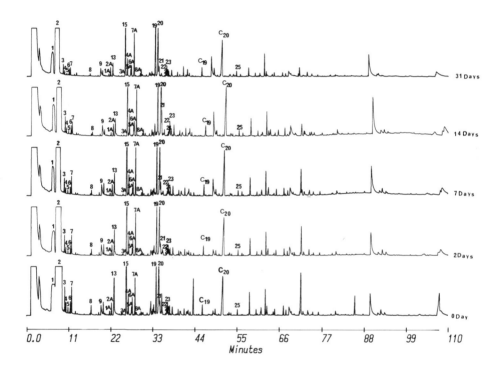

Figure 3.   Gas chromatogram of volatile components remained
            in orange juice after storage at 25 C for 0, 2, 7,
            14 and 31 days.

Table I.Volatile Compounds Eluted from Inner Package After
Storage at 4 C for 2, 14, and 31 days

4 C Peak No.	R.I.	Compound	date 2	14	31
			(A sample / A int std)		
1.	1146	beta-pinene	–	0.16	1.06
2.	1184	limonene	0.55	52.59	242.03
3.	1225	4-carene	0.06	0.08	0.35
4.	1232	trans-ocimene	–	0.03	0.11
5.	1238	p-cymene	–	0.03	0.11
6.	1258	monoterpene mw.136	–	0.08	0.27
7.	1266	octanal	–	0.03	0.10
8.	1361	nonanal	0.05	0.15	0.19
9.	1425	1-methyl-4-(2-methyl-oxiranyl)-7-oxabicyclo [4.1.0]heptane	0.13	0.33	0.63
10.	1454	citronellal	0.03	0.09	0.05
11.	1460	sesquiterpene mw.204	0.05	0.14	0.17
12.	1463	sesquiterpene mw.204	0.15	0.49	0.52
13.	1473	decanal	0.48	2.11	2.46
14.	1512	unknown	0.12	0.31	4.82
15.	1530	linalool	0.06	0.15	0.45
16.	1553	sesquiterpene mw.204	0.33	1.66	1.20
17.	1569	unknown	0.23	0.22	0.30
18.	1573	unknown	0.05	0.27	0.34
19.	1665	alpha-terpineol	0.15	0.41	0.76
20	1673	valencene?	3.05	10.04	9.50
21.	1683	neral	0.10	0.19	0.13
22.	1709	sesquiterpene	0.21	0.58	0.51
23.	1714	sesquiterpene	0.33	0.81	0.84
24.	1720	perillaldehyde	0.10	0.17	0.14
25.		ethyl cinnamate(int std)	1.00	1.00	1.00
		total	6.23	71.12	267.04

Table II.Volatile Compounds Eluted from Inner Package After
Storage at 25 C for 2, 7, 14, and 31 days

25 C Peak No.	R.I.	Compound	date			
			2	7	14	31
			(A sample / A int std)			
1.	1146	beta-pinene	0.24	1.18	4.1	3.43
2.	1184	limonene	30.60	242.40	568.54	625.56
3.	1225	4-carene	0.08	0.38	0.81	1.01
4.	1232	trans-ocimene	0.02	0.11	0.19	0.21
5.	1238	p-cymene	0.07	0.07	0.10	0.18
6.	1258	monoterpene mw.136	0.07	0.29	0.54	0.65
7.	1266	octanal	0.04	0.12	0.21	0.16
8.	1361	nonanal	0.15	0.20	0.38	0.29
9.	1425	1-methyl-4-(2-methyl- oxiranyl)-7-oxabicyclo [4.1.0]heptane	0.71	0.30	0.53	1.38
10.	1454	citronellal	0.17	0.12	0.12	0.15
11.	1460	sesquiterpene mw.204	0.32	0.17	0.31	0.30
12.	1463	sesquiterpene mw.204	-	0.39	0.85	1.37
13.	1473	decanal	2.51	2.27	2.90	2.34
14.	1512	unknown	1.88	0.14	0.34	9.12
15.	1530	linalool	0.34	0.63	1.24	1.18
16.	1553	sesquiterpene mw.204	0.68	0.83	2.42	3.39
17.	1569	unknown	0.44	0.58	0.71	0.82
18.	1573	unknown	0.21	0.14	0.31	0.30
19.	1665	alpha-terpineol	0.64	0.52	1.14	2.02
20	1673	valencene?	5.52	6.19	14.74	21.96
21.	1683	neral	0.51	0.10	0.30	0.37
22.	1709	sesquiterpene	0.38	0.39	0.91	1.31
23.	1714	sesquiterpene	0.54	0.66	1.3	1.95
24.	1720	perillaldehyde	0.16	0.15	0.31	0.41
25.		ethyl cinnamate(int std)	1.00	1.00	1.00	1.00
		total	46.28	258.33	603.30	679.86

Table III.Volatile Compounds Eluted from Inner Package After
Storage at 37 C for 2, 7, 14, and 31 days

37 C Peak No.	R.I.	Compound	date			
			2	7	14	31
			(A sample / A int std)			
1.	1146	beta-pinene	0.02	0.46	0.50	1.54
2.	1184	limonene	1.88	122.13	104.50	350.29
3.	1225	4-carene	0.05	0.19	0.19	0.62
4.	1232	trans-ocimene	0.04	0.04	0.04	0.12
5.	1238	p-cymene	0.05	0.02	0.05	0.11
6.	1258	monoterpene mw.136	0.18	0.19	0.18	0.65
7.	1266	octanal	0.04	0.05	0.04	0.06
8.	1361	nonanal	0.06	0.16	0.16	0.12
9.	1425	1-methyl-4-(2-methyl-oxiranyl)-7-oxabicyclo[4.1.0]heptane	0.63	0.28	0.54	1.36
10.	1454	citronellal	0.04	0.09	0.12	0.14
11.	1460	sesquiterpene mw.204	0.26	0.16	0.17	0.32
12.	1463	sesquiterpene mw.204	-	0.57	1.00	1.59
13.	1473	decanal	0.67	2.15	1.57	1.44
14.	1512	unknown	1.13	0.15	1.24	16.72
15.	1530	linalool	0.03	0.53	0.37	0.67
16.	1553	sesquiterpene mw.204	0.54	1.14	2.63	3.95
17.	1569	unknown	0.19	0.62	0.47	0.71
18.	1573	unknown	0.10	0.21	0.28	0.10
19.	1665	alpha-terpineol	0.22	0.83	1.74	2.87
20	1673	valencene?	5.25	10.36	16.64	24.44
21.	1683	neral	0.09	0.16	0.28	0.29
22.	1709	sesquiterpene	0.33	0.57	0.99	1.55
23.	1714	sesquiterpene	0.53	1.00	1.64	2.50
24.	1720	perillaldehyde	0.12	0.22	0.49	0.69
25.		ethyl cinnamate(int std)	1.00	1.00	1.00	1.00
		total	12.45	142.28	135.83	412.85

Table IV.Volatile Compounds of Aseptic Orange Juice Stored
at 4 C for 2, 7, 14, and 31 days

Peak R.I.	Compounds	Date				
No. CW-20M		0	2	7	14	31
			(A sample / A int std)			
1. 1146	beta-pinene	136.02	104.37	102.69	79.89	75.75
2. 1184	limonene	8813	6460	6076	4855	4509
3. 1225	4-carene	10.23	9.39	9.57	7.92	6.75
4. 1232	trans-ocimene	3.21	2.58	2.67	1.92	1.89
5. 1238	p-cymene	2.13	1.59	1.65	1.62	1.26
6. 1258	monoterpene mw.136	5.28	3.90	3.90	3.54	2.82
7. 1266	octanal	14.49	11.01	10.20	10.35	9.45
8. 1361	nonanal	5.79	3.81	3.27	3.30	2.76
9. 1425	1-methyl-4-(2-methyl-oxiranyl)-7-oxabicyclo [4.1.0]heptane	11.79	16.41	14.55	14.01	9.96
1A. 1457	unk cpd	2.19	1.53	1.47	1.65	1.38
2A. 1463	unk sesquiterpene m.w.204	6.48	6.48	6.90	7.17	5.31
13. 1473	decanal	35.25	25.80	19.86	20.94	15.54
3A. 1511	unk cpd	1.41	1.65	0.96	0.99	0.54
15. 1530	linalool	98.25	93.60	88.32	101.70	94.86
4A. 1538	1-octanol	12.63	11.91	11.67	13.77	12.57
5A. 1553	unk sesquiterpene m.w.204	4.98	4.95	4.98	4.92	5.04
6A. 1557	unk sesquiterpene m.w.204	8.91	8.58	10.11	9.30	6.81
7A. 1570	1-terpinen-4-ol	21.99	25.08	26.79	31.41	29.46
8A. 1575	dihydro-4,5-dimethyl-2(3H)-furanone ?	2.52	1.71	0.81	0.99	0.84
19. 1665	alpha-terpineol	39.27	37.20	40.38	47.01	42.03
20. 1673	valencene?	60.33	60.21	71.64	68.64	49.80
21. 1683	neral	7.68	7.20	13.08	12.78	5.37
22. 1709	sesquiterpene	5.52	5.55	4.38	4.41	5.64
23. 1714	sesquiterpene	6.78	6.51	7.59	6.81	5.22
24. 1720	perillaldehyde	1.98	1.89	2.01	1.68	1.50
25.	ethyl cinnamate (int std)	1.00	1.00	1.00	1.00	1.00
	total	9318	6913	6535	5312	4902

Table V.Volatile Compounds of Aseptic Orange Juice Stored
at 25 C for 2, 7, 14, and 31 days

					Date		
Peak No.	R.I. CW-20M	Compounds	0	2	7	14	31
			(A sample / A int std)				
1.	1146	beta-pinene	136.02	87.00	72.57	76.53	68.31
2.	1184	limonene	8813	5431	4369	4822	4011
3.	1225	4-carene	10.23	9.60	8.04	8.31	5.04
4.	1232	trans-ocimene	3.21	2.61	2.25	2.13	1.53
5.	1238	p-cymene	2.13	1.47	1.26	1.35	0.84
6.	1258	monoterpene mw.136	5.28	3.81	3.12	3.57	2.82
7.	1266	octanal	14.49	9.75	8.10	5.37	3.78
8.	1361	nonanal	5.79	2.97	2.34	1.80	1.35
9.	1425	1-methyl-4-(2-methyl-oxiranyl)-7-oxabicyclo [4.1.0]heptane	11.79	14.13	12.93	11.79	12.00
1A.	1457	unk cpd	2.19	1.44	1.41	1.65	1.14
2A.	1463	unk sesquiterpene m.w.204	6.48	5.91	5.88	6.66	4.68
13.	1473	decanal	35.25	16.62	14.40	10.26	6.84
3A.	1511	unk cpd	1.41	1.32	0.78	0.57	0.39
15.	1530	linalool	98.25	92.19	87.81	75.87	69.51
4A.	1538	1-octanol	12.63	12.3	12.57	11.40	10.17
5A.	1553	unk sesquiterpene m.w.204	4.98	4.65	4.41	4.92	4.44
6A.	1557	unk sesquiterpene m.w.204	8.91	7.95	7.20	8.55	6.12
7A.	1570	1-terpinen-4-ol	21.99	25.71	26.58	25.26	24.81
8A.	1575	dihydro-4,5-dimethyl-2(3H -furanone ?	2.52	0.84	0.39	0.57	0.60
19.	1665	alpha-terpineol	39.27	41.61	45.06	47.67	54.96
20.	1673	valencene?	60.33	53.61	51.36	76.8	51.99
21.	1683	neral	7.68	6.48	5.82	10.95	4.95
22.	1709	sesquiterpene	5.52	5.49	3.57	4.71	5.04
23.	1714	sesquiterpene	6.78	6.27	6.12	9.15	5.49
24.	1720	perillaldehyde	1.98	1.83	1.68	1.11	1.53
25.		ethyl cinnamate (int std)	1.00	1.00	1.00	1.00	1.00
		total	9318	5847	4755	5226	4359

Table VI.Volatile Compounds of Aseptic Orange Juice Stored
at 37 C for 2, 7, 14, and 31 days

Peak No.	R.I. CW-20M	Compounds	0	2	7	14	31
			(A sample / A int std)				
1.	1146	beta-pinene	136.02	88.50	76.05	69.57	29.34
2.	1184	limonene	8813	5587	4356	3993	1770
3.	1225	4-carene	10.23	9.27	7.98	6.75	3.63
4.	1232	trans-ocimene	3.21	2.43	1.95	1.98	1.35
5.	1238	p-cymene	2.13	1.35	1.23	1.17	0.93
6.	1258	monoterpene mw.136	5.28	3.90	3.75	3.90	2.91
7.	1266	octanal	14.49	9.54	6.96	4.56	2.04
8.	1361	nonanal	5.79	2.70	1.95	1.35	0.60
9.	1425	1-methyl-4-(2-methyl-oxiranyl)-7-oxabicyclo[4.1.0]heptane	11.79	14.94	17.85	20.31	18.00
1A.	1457	unk cpd	2.19	1.47	1.29	1.17	0.78
2A.	1463	unk sesquiterpene m.w.204	6.48	6.27	5.64	5.58	3.42
13.	1473	decanal	35.25	17.76	9.33	7.86	3.06
3A.	1511	unk cpd	1.41	1.05	1.11	0.96	0.27
15.	1530	linalool	98.25	88.50	84.99	73.44	47.73
4A.	1538	1-octanol	12.63	11.91	12.24	11.70	10.41
5A.	1553	unk sesquiterpene m.w.204	4.98	4.65	3.93	3.75	3.72
6A.	1557	unk sesquiterpene m.w.204	8.91	7.74	6.81	6.42	3.69
7A.	1570	1-terpinen-4-ol	21.99	26.04	26.58	26.31	22.38
8A.	1575	dihydro-4,5-dimethyl-2(3H)-furanone ?	2.52	0.93	0.72	0.54	0.42
19.	1665	alpha-terpineol	39.27	43.29	60.30	74.85	95.79
20.	1673	valencene?	60.33	56.40	50.91	53.19	30.96
21.	1683	neral	7.68	6.51	11.19	10.29	3.24
22.	1709	sesquiterpene	5.52	3.63	3.57	3.66	2.52
23.	1714	sesquiterpene	6.78	6.48	6.06	6.30	4.32
24.	1720	perillaldehyde	1.98	1.71	1.53	1.29	1.08
25.		ethyl cinnamate (int std)	1.00	1.00	1.00	1.00	1.00
		total	9318	6003	4760	4390	2062

flavor component derived from limonene, increased significantly in juice stored at higher temperature and longer time.

Figure 4 shows the summation of total area of all treatments shown in Tables I to VI. It is very clear that only trace amounts of volatile components were absorbed by the inner packaging material when aseptic packages were stored at the lower temperature. However, large amounts of volatile components were absorbed when the aseptic packages were stored at the higher temperature. By comparing the rate of absorption and the rate of loss (volatile components remained in juice), it is interesting to note that oxidation or thermal induced degradation of volatile components may have significant effects in determining the quality of aseptic orange juice.

Table VII shows the sensory results of all treatments in the present study. Aseptic orange juice stored at 4 C showed highest score throughout the whole storage period. By comparison, juice stored at 37 C for two weeks became unacceptable, and significant difference was noted after one week. For juice stored at 25 C, detectable difference was noted after two weeks, but the sensory quality was still above acceptable level after 4 weeks. Similar results were reported by Marcy et al. (21).

Table VII.Sensory Scores of Aseptic Orange Juice Stored at Different Temperatures

Temp	Days				
	0	2	7	14	28
4 C	6.00	6.182a	6.29a	6.54a	6.29a
25 C	6.00	6.091a	5.63a	5.49b	4.46b
37 C	6.00	5.958a	3.58b	4.21c	3.42c

NOTE :Duncan's multiple range test, sensory values within the same column with the same letter are not significant different from each other for $P < 0.05$.
Nine point hedonic scale: 1=dislike extremely; 9=like extremely.

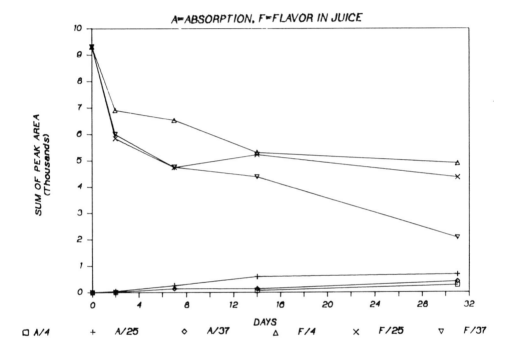

Figure 4.   The summation of total peak area of all treatments
shown in Tables I to IV.   A/4, A/25 and A/37 stand
for summation of volatile components absorbed by
inner packages that were stored at 4 C, 25 C and
37 C, for 2, 7, 14, and 31 days.   F/4, F/25 and
F/37 stand for the summation of volatile
components remaining in juice after storage at 4
C, 25 C and 37 C, for 2, 7, 14, and 31 days.

Conclusion

Quantitative analysis of the volatile components of aseptic orange juice absorbed by the inner packaging material and those remaining in juice, in conjuction with sensory analysis, have lead to the conclusion that the storage temperature and time will have a profound effect upon the quality of aseptic orange juice. Absorption of volatile components by the inner packaging material may be important but only to a certain extent.

Acknowledgments

The technical assistance of Dr. T.-C. Chang, Mr. Steve Ding, Mr. Y.-H. Kuo and Ms. B.-S. Kao, Food Industry Research and Development Institute, is appreciated. This research was supported by Department of Economics, Republic of China.

Literature Cited

1. Durr, P., Schobinger, U. 1981, The contribution of some volatiles to the sensory quality of apple and orange juice odour. in "Flavor '81" P. Schreier(ed), Walter de Gruyter, pp179-192.
2. Graumlich, t. R., Marcy, J. E., Adams, J. P. 1986 J. Agric. Food Chem. 34, 402.
3. Sizer, C.E., Waugh, P.L., Edstam, S., and Ackermann, P. 1988, Food Technol. 42: 152.
4. Moshonas, M.G. and Shaw, P.E. 1989, J. Agric. Food Chem. 37: 157.
5. Marshall, M.R., Adams, J.P., and Williams, J.W. 1985, Proceedings of third international conference and exhibition on aseptic package aseptipak '85 p: 299.
6. Ackermann, P.W. and Wartenberg, E.W. 1986, Shelf-life of juices: a comparison between different packages. Report of 19th Symposium of international federation of fruit juice producers. Den Haag, p.143.
7. Mannheim, C. H., Miltz, J., Passy, N. 1988, ACS Symp. Ser. 365 pp 68-82.
8. Hirose, K., Harte, B. R., Giacin, J. R., Miltz, J., Stine, C. 1988 ACS Symp. Ser. 365 pp 28-41.
9. Halek, G. W., Meyers, M. A., 1989, Packag. Technol. Sci. 2, 141-146.
10. Imai, T., Harte, B.R., Giaicin, J.R. 1990 J. Food Sci. 55(1), 158.
11. Chen, C.-C., Lin, S.-J., Chen, S.-Y., Chen, C.-C., Zen, S.-M., Jeng, J.-G. 1990, Development and evaluation of a laboratory scale aseptic processing and packaging system. Research Report. Food Industry R&D Institute, Hsinchu, Taiwan, ROC.
12. Marsili, R., Kilmer, G., and Miller, N. 1989, LC.GC 7: 778.
13. Shimoda, M., Nitada, T., Kadota, N., Ohta, H., Suetsuna, K., Osajima, Y. 1984, Nippon Shokuhin Kogyo Gakkaishi. 30(11), 697-703.
14. Majlat, P. Erdos, Z. Takacs, J. J. Chromatogr. 1974, 91, 89.
15. SAS, 1985, "SAS User's Guide:Statistics."Version 5 ed., SAS Institute, Inc., Cary, NC.

16. Potter, R. H., Bertels, J.R., Sinki, G., 1985, Proceedings of
    third international conference and exhibition on aseptic package
    aseptipak '85: pp 313-338
17. Fleisher, A., Biza, G., Secord, N., Dono, J. 1987, Perfum &
    Flavor. 12(2), 57.
18. Moshonas ,M.G. and Shaw, P.E. 1987, J. Agric. Food Chem. 35:
    161.
19. Swaine,R.L., Swaine, R.L.Jr. 1988, Perf. & Flav. 13(6): 1.
20. MacLeod, A. J., MacLeod, G., Subramanian, G. 1988, Phytochem.
    27(7), 2185.
21. Marcy, J. E., Hansen, A. P., Graumlich, T.R. 1989, J. Food
    Sci. 54, 227.

RECEIVED March 11, 1991

Chapter 17

Sorption of Flavor Compounds by Polypropylene

D. K. Arora[1], A. P. Hansen, and M. S. Armagost

Department of Food Science, North Carolina State University, Raleigh, NC 27695−7624

The sorption of aldehydes (C_7-C_9), methyl ketones (C_7-C_{10}), methyl esters (C_7-C_9), sulfur compounds and alkylpyrazines by polypropylene (PP) was investigated. PP discs (163.9 cm^2) were immersed in a buffer solution containing a flavor compound (200 ppm) for 24 hr at room temperature. The flavor compound remaining in the buffer and sorbed by the PP was quantitated using GC. The percent of flavor compounds sorbed by PP was: aldehydes from 17.2% to 63.8%, methyl ketones from 4.2% to 73.7%, methyl esters from 13.0% to 79.4%, and sulfur compounds from 38.7% to 48.6%. Alkylpyrazines were sorbed to an insignificant level (1.0-2.2%). For each group of flavor compounds, the sorption increased as the carbon number increased.

The use of plastics in the packaging of food products and pharmaceuticals has increased considerably over the last several years. In terms of dollar value, an increase of 56% is estimated for plastics from 1984 to 1990 (1). Bob Schwier of Schotland Business Research predicted that the use of coextruded plastic barrier materials in food processing world wide would exceed $4 billion by 1990 (2). Many recent developments in food packaging are related to using the barrier properties of plastics to extend shelf life.

Recently, there has been a surge in aseptic processing and packaging of food and pharmaceutical products for quality, convenience and longer shelf life. Fruit juices and drinks had the most impressive growth

[1]Current address: Engineering Resources, Inc., Fayetteville, AR 72702

0097−6156/91/0473−0203$06.00/0
© 1991 American Chemical Society

in aseptic packaging. Other aseptic products include
tomato sauce, apple sauce, gravies, cheese dips, soups,
flavored milks, puddings and yogurt products. Currently,
high acid liquid foods account for 90% of the aseptic
market (3). When sterilization techniques for
particulates in low acid foods are commercially
feasible, a large increase in aseptic packaging is
anticipated. It is estimated that in the year 2000 the
U.S. market may have a volume of over 15 billion aseptic
packages (4).
    Aseptic foods are packaged in a variety of
polymeric materials. The aseptic packages are formed
from thermoplastics or made by combining thermoplastics
with paperboard and metal foil. Plastics used in aseptic
packages may be single polymers or coextruded polymers
designed to improve barrier properties. Most aseptic
food containers have an inner plastic liner for heat
sealing. Plastics are not as inert as glass and metals,
and can interact with foods and pharmaceuticals packaged
in the container. Various interactions between plastics
and food products, such as permeability, migration and
sorption, have been described (1). The sorption of
flavor compounds by polymeric materials refers to the
scalping of flavor and aroma components of foods by
plastics. Scalping of flavor compounds has been shown to
influence the sensory quality and acceptability of food
products (5,6).
    The shelf stability of fruit juices and fruit
drinks in aseptic packages has been documented (5-11).
The PE liner in brick type aseptic containers scalps
d-limonene, neral, geranial, octanal and decanal from
juice products (Adams, J.P., International Paper Co.,
personal communication, 1988.). As a result, the
packaged fruit juice lacks flavor notes characteristic
of fresh juice. Using headspace chromatography, Abbe et
al. (12) attempted to "fingerprint" key flavor compounds
sorbed by the aseptic package containing a juice
product. The package that contained juice showed extra
peaks in the chromatogram as compared to the control
container, which did not contain juice. The extra peaks
represented the uptake of flavor compounds by the
package.
    Salame (Salame, M., Paper presented at Bev-Pak '88,
unpublished data, 1988.) reported an average of 62% and
37% sorption of non-polar and polar organics,
respectively, by PP. Polypropylene was ranked second
after PE in the quantity of flavor compounds sorbed
(Salame, M., Paper presented at Bev-Pak '88, unpublished
data, 1988.). Juice packaged in coextruded polypropylene
containers, consisting of PP outer and inner layers
surrounding an internal ethylene vinyl alcohol layer,
showed a marked decrease in d-limonene content as
compared to juice packaged in foil containers due to
sorption (13). A layer of Saran has been suggested with

PP and HDPE to reduce the sorption of d-limonene (14).
The polarity and size of the permeant, the area of
polymer, the thickness of film, the storage temperature
and the relative humidity affect the sorption of flavor
compounds by polymers (15).
The sorption of five classes of flavor compounds
(aldehydes, methyl ketones, methyl esters,
alkylpyrazines and sulfur compounds) representing a wide
variety of food products, such as dairy, fruit,
vegetable and meat products by PP at room temperature
was investigated.

Materials and Methods

Materials. The following flavor compounds were used:
aldehydes (heptanal, octanal and nonanal), methyl
ketones (2-heptanone, 2-octanone, 2-nonanone and
2-decanone), methyl esters (methyl hexanoate, methyl
heptanoate and methyl octanoate), alkylpyrazines
(methyl- and ethylpyrazine) and sulfur compounds
(2-methylthiophene and benzyl methyl sulfide). These
chemicals were obtained from Aldrich Chemical Company,
Milwaukee, WI. The polypropylene (PP) sheet (0.3 mm
thick) was obtained from Kraft General Foods, Inc.,
Glenview, IL.

Sorption Apparatus and Method. The PP was cut into discs
of 2.7 cm diameter. Fourteen discs (163.9 cm^2, surface
area) were mounted onto a silanized stainless steel wire
alternating with silanized glass beads (4 mm, #4000)
(Figure 1). The stainless steel wire with PP discs and
glass beads was inserted into a silanized screw tight
glass bottle. The screw cap was lined with tin foil
(Fisher Scientific, Raleigh, NC) to prevent scalping of
the flavor compound by the cap. Each of the glass
bottles contained a 2% alcoholic buffer (100 mL) of
simulated milk ultrafiltrate (SMUF) (16) and a flavor
compound (200 ppm). The bottles were placed on a shaker
for uniform dispersal of the flavor compound for 24
hours at room temperature (24°C). Two sets of controls
(buffer solution containing only the flavor compound and
buffer solution containing the flavor compound, wire
and glass beads) were placed on the shaker along with
the bottles containing the PP samples.

Flavor Compound Extraction. The following extraction
method was adopted for quantitating flavor compound
sorption.   After 24 hours, the PP sample was removed
from the bottle. Sodium chloride (10 g) was added to the
buffer solution to break the emulsion. The flavor
compound in the buffer solution was extracted with
methylene chloride (5 mL x 4). The methylene chloride
from each extract was pooled and treated with anhydrous

Na_2SO_4 to remove all traces of water before analysis by GC.

With the PP discs still mounted on the wire, the sample was rinsed with distilled water (100 mL) to remove the loosely adhered flavor compound from the surface of the discs, wire and glass beads. The rinse water was analyzed in the same manner as the buffer. After rinsing, the discs were extracted with methylene chloride (5 mL x 3). The methylene chloride from the three extractions was pooled and treated with anhydrous Na_2SO_4 to remove all traces of water before analysis by GC.

The buffer solution in the controls was extracted in the same manner as the buffer solution containing the PP discs. For controls containing the wire and glass beads, the wire and beads were extracted for adhered compounds in the same manner as the discs.

The quantitation of flavor compounds extracted from the PP and from the buffer were performed in triplicate. The quantity of flavor compound extracted was determined using peak areas. Calibration curves were generated using peak areas from known concentrations of flavor compounds. Recovery studies were performed to determine the efficiency of the extraction method. The statistical analyses were performed using General Linear Models Procedure by Waller-Duncan (17).

Gas Chromatography. Gas chromatographic analysis was performed using a Varian model 3700 gas chromatograph with a FID detector connected to a Hewlett Packard model 3390A integrator. The column oven temperature was programmed as follows: 120°C for 1 min initial hold, 120°C-240°C at 5°C/min, and 240°C for 10 min final hold. Injector and detector temperatures were 240°C and 250°C, respectively. A Supelcowax™ 10, (Supelco, Inc., Bellefonte, PA) fused silica capillary column (30 m x 0.53 mm and 1.0 μm film thickness) was used.

Results and Discussion

The percent of flavor compounds sorbed by the PP samples and the quantity of flavor compounds extracted and quantitated from the PP, buffer and rinse water are presented in Table I.

Recovery Study. The percent recovery for each flavor compound was calculated by comparing the quantity recovered from the two sets of controls with the quantity initially placed in the buffer. There was no significant difference (P<0.05) between the control that did not contain wire and beads and the control that contained wire and beads. The recovery of aldehydes, methyl ketones, methyl esters, sulfur compounds and alkylpyrazines was 86.3%, 91.4%, 92.5%, 96.4% and 89.7%, respectively.

Aldehydes. An average of 15.3% (+3.9) of the aldehydes was recovered from the rinse water. The total amount of aldehydes recovered after extracting the discs, buffer solution, and rinse water was about 88% of the control for all the aldehydes.

Methyl Ketones. An average of 7.2% (+2.6) of the ketones was recovered from the rinse water. The total amount of ketones recovered after extracting the discs, buffer and rinse water was about 95% of the control for all the ketones.

Methyl Esters. An average of 8.0% (+4.8) of the esters was recovered from the rinse water. The total amount of methyl esters recovered after extracting the PP, buffer and rinse water was about 96% of the control of all the methyl esters.

Sulfur Compounds. An average of 0.4% (+0.1) of sulfur compounds were recovered from the rinse water. The average recovery for sulfur compounds after extracting the PP, buffer and rinse water was about 97% of the control of all sulfur compounds.

Alkylpyrazines. Methyl- and ethylpyrazines were sorbed to insignificant levels of 1.0% and 2.2%, respectively.

Distribution Ratios. The distribution ratio is defined as the ratio of the amount of flavor compound sorbed into the PP to the amount remaining in the buffer. Figure 2 shows the distribution ratio of flavor compound groups (aldehydes, methyl ketones and methyl esters) vs. carbon number. As the carbon number of the flavor compound increases, an increase in the distribution ratio was observed. The higher hydrophobicity of the longer chain flavor compounds may cause increased sorption into PP discs. The greater sorption of longer chain flavor compounds has been previously reported (18-20). The following distribution ratios of (C_9) flavor compounds was observed: methyl ester (5.00), aldehyde (2.56), and methyl ketones (0.63). These ratios suggest that straight chain aldehydes and methyl esters were sorbed to a greater extent ($P<0.05$) than branched chain methyl ketones.

For aldehydes, methyl ketones and methyl esters, the increased sorption of higher molecular weight flavor compounds agreed with literature values. Landois-Garza and Hotchkiss (15) have reported higher permeation for higher molecular weight esters into polyvinyl alcohol. Zobel (21) also observed a similar trend for

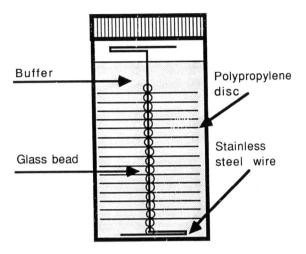

Figure 1.   Sorption assembly for polypropylene.

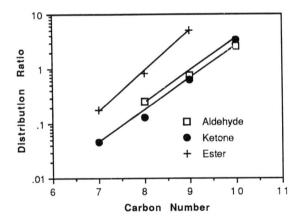

Figure 2.   Distribution ratios for aldehydes, methyl ketones and methyl esters.

Table I

Percentage of Flavor Compounds Sorbed by Polypropylene (PP) and Quantity of Flavor Compounds Extracted from PP, Buffer and Rinse Water

Compounds	Percent Sorbed by PP	Quantity (mM) of Flavor Compounds		
		PP	Buffer	Rinse Water
Aldehydes				
Heptanal	17.2	0.79 (±0.04)	3.06 (±0.04)	0.75 (±0.10)
Octanal	35.4	1.52 (±0.05)	1.97 (±0.04)	0.80 (±0.15)
Nonanal	63.8	2.11 (±0.03)	0.83 (±0.01)	0.37 (±0.12)
Methyl Ketones				
2-Heptanone	4.2	0.21 (±0.01)	4.42 (±0.04)	0.32 (±0.05)
2-Octanone	10.5	0.60 (±0.05)	4.50 (±0.17)	0.61 (±0.10)
2-Nonanone	35.7	1.73 (±0.08)	2.74 (±0.33)	0.38 (±0.07)
2-Decanone	73.7	2.49 (±0.43)	0.73 (±0.28)	0.16 (±0.05)
Methyl Esters				
Methyl Hexanoate	13.0	0.67 (±0.01)	3.78 (±0.06)	0.70 (±0.04)
Methyl Heptanoate	42.8	1.79 (±0.32)	2.15 (±0.12)	0.24 (±0.05)
Methyl Octanoate	79.4	3.39 (±0.14)	0.68 (±0.04)	0.20 (±0.07)
Sulfur Compounds				
2-Methylthiophene	48.6	4.02 (±0.38)	4.23 (±0.06)	0.03 (±0.01)
Benzyl Methyl Sulfide	38.7	2.11 (±0.04)	3.32 (±0.11)	0.02 (±0.01)
Alkylpyrazines				
Methylpyrazine	1.0	0.06 (±0.02)	5.90 (±0.16)	0.25 (±0.05)
Ethylpyrazine	2.2	0.14 (±0.01)	6.15 (±0.14)	0.22 (±0.07)

permeability of esters through polypropylene. The permeability of a compound through a polymer is a function of diffusion and solubility. As the molecular weight of the compound increases, diffusivity decreases, possibly due to steric hinderance, however, solubility increases to greater extent resulting in higher sorption and permeation of a flavor compound (15).

A few treatments to polymer surfaces and the judicious selection of polymers, such as use of crystalline polymers (18), use of polymer blends (22), fluorination of polymer surfaces (23,24) and inert coatings (silica) on the polymer surfaces (25) have been proposed in the literature to circumvent the scalping of flavor compounds from the food products.

Packages for aseptically processed foods and pharmaceuticals consist of a variety of polymers. Sorption of flavor compounds in these packages may result in products with an imbalance of flavor and aroma. Determining the quantity of flavor compounds sorbed will help food processors in selecting packaging materials and food ingredients for stable flavor and aroma.

Acknowledgments. Paper number FSR91-04 of the journal series of the Department of Food Science, North Carolina State University, Raleigh, NC 27695-7624. The use of trade names in this publication does not imply endorsement by the North Carolina Agricultural Research Service of the products named, nor criticism of similar products not mentioned. The research reported in this publication was funded by the Center for Aseptic Processing and Packaging Studies (CAPPS) and by the North Carolina Agricultural Research Service. The authors would like to thank J.J. Heinis for his assistance in preparation of this manuscript.

Literature Cited

1. Hotchkiss, J.H.   In Food and Packaging Interactions; Hotchkiss J.H., Ed.; American Chem. Soc.: Washington, DC, 1988; pp. 1-10.
2. Brody, A.L. Food Eng. 1987, 59(4), 49-52.
3. Judy, R. Food Proc. 1988, 49(10), 35-53.
4. Sacharow, S. Prepared Foods 1986, 155(1), 31-32.
5. Moshonas, M.G. and Shaw, P.E. J. Agric. Food Chem. 1989, 37, 157-161.
6. Graumlich, T.R., Marcy, J.E. and Adams, J.P. J. Agric. Food Chem. 1986, 34, 402-405.
7. Tillotson, J.E. Food Tech. 1984, 38(3), 63-66.
8. Imai, T., Harte, B.R. and Giacin, J.R. J. Food Sci. 1990, 55, 158-161.
9. Mannheim, C.H., Miltz, J. and Letzer, A. J. Food Sci. 1987, 52, 737-740.

10. Moshonas, M.G. and Shaw, P.E. In Frontiers of Flavors; Charalambous, G., Ed.; Proceedings of the Fifth International Flavor Conference; Porto Karras: Chalkidiki, Greece, 1987; pp. 133-145.
11. Kwapong, O.Y. and Hotchkiss, J.H. J. Food Sci. 1987, 52, 761-763, 785.
12. Abbe, S., Bassett, B.M. and Collier, J.C. Polymers, Laminations, and Coating Conference; TAPPI Conference, 1985; p.411-416.
13. Mottar, J. 1990. FSTA. 1990, 22, 61.
14. DeLassus, P.T. Polymer, Laminations and Coatings Conference; TAPPI Proceeding, 1985; p. 445-450.
15. Landois-Garza, J. and Hotchkiss, J.H. Food Eng. 1987, 59, 39-42.
16. Sood, S.M., Sindhu, K.S. and Dewan, R.K. Milchwassen-schaft. 1976, 31, 470.
17. SAS Institute Inc. SAS User's Guide: Statistics, 1982 Edition; SAS Institure Inc.: Cary, NC, 1982.
18. Shimoda, M., Ikegami, T. and Osajima, Y. J. Sci. Food Agric. 1988, 42, 157-163.
19. Krizen, T.D., Cuburn, J.C. and Blatz, P.S. In Barrier Polymers and Structures; Koros, W.J., Ed.; ACS symposium series 423; American Chemical Society: Washington, DC, 1990; p. 111-125.
20. Strandburg, G., DeLassus, P.T. and Howell, B.A. In Barrier Polymers and Structures; Koros, W.J., Ed.; ACS symposium series 423; American Chemical Society: Washington, DC, 1990; p. 333-350.
21. Zobel, M.G.R. 1985. Polymer Testing. 1985, 5, 153-165.
22. Subramanian, P.M. In Barrier Polymers and Structures; Koros, W.J., Ed.; ACS symposium series 423; American Chemical Society: Washington, DC, 1990; pp. 252-265.
23. Hobbs, J.P., Anand, M. and Campion, B.A. In Barrier Polymers and Structures; Koros, W.J., Ed.; ACS symposium series 423; American Chemical Society: Washington, DC, 1990; pp. 280-294.
24. Mehta, R.K. and Buck, D.M. Packaging. 1988 33(10), 100-101.
25. Charara, Z.N. Evaluation of Orange Flavor Absorption into Various Polymeric Packaging Materials; M.S. Thesis, University of Florida, 1987.

RECEIVED March 1, 1991

Chapter 18

Sorption Behavior of Citrus-Flavor Compounds in Polyethylenes and Polypropylenes

Effects of Permeant Functional Groups and Polymer Structure

George W. Halek and Joseph P. Luttmann

Polymeric Materials Research Laboratory, Food Science Department, Rutgers University—The State University of New Jersey, P.O. Box 231, New Brunswick, NJ 08903

An understanding of the interactions between food ingredients and polymeric packaging materials requires knowledge of the chemical and physical structures of the food ingredients and of the polymers (1). This paper reports the behavior of citrus compounds with varied functional groups in model water solutions with polyolefins of three structural types: Unbranched high density polyethylene, branched low density polyethylene and two stereoregular polypropylenes. Included in the study are the partitioning behavior of a mixture of seven citrus flavor compounds and a comparison of the partition coefficients and Gibbs free energy changes of limonene and carvone representing compounds of similar structure but with different polarities. Mechanistic implications are drawn from the data on their interactions with these structurally different polyolefins.

A number of studies have shown that flavor ingredients in orange juice are sorbed by polymeric packaging materials. Dürr and Schobinger (2) and Dürr et al. (3), reported that orange juice lost flavor when stored in polyethylene containers at room temperature for three months. Marshall et al. (4), and Mannheim et al. (5), reported that d-limonene was sorbed into the polyethylene lining of a paperboard package. Kwapong and Hotchkiss (6) and Hirose et al. (7) reported that significant citrus flavor absorption occurred into polyethylene within a few days. Halek and Meyers (8) showed that the sorption of a number of citrus flavor compounds by low density polyethylene was almost instantaneous and partitioning depended on the polarity of the compounds. Imai et al. (9) , reported that low density polyethylene sorbed more limonene from orange juice than did ethylene/vinyl alcohol copolymer of high ethylene content (EVOH) or a co-polyester (Co-Pet). Thus, it is well established that sorption does occur, but very little data have been reported on the effects of the structure and polarity of the flavors and of the polymers on the sorption

0097–6156/91/0473–0212$06.00/0

behaviors and even less has appeared on interpretation of the reasons for the behaviors. Such interpretation is important because it is known that the typical orange aroma is attributed to a mixture of acids, alcohols, aldehydes, esters, hydrocarbons, ketones and other components, and that the correct proportions of the different components are critical to fresh orange flavor *(10)*. If the rates of sorption and ultimate equilibrium partitioning of the chemically different flavor ingredients in a plastic package were different, the flavor profile would change over time.   This could present a problem in the practice of compensating for flavor ingredient changes by adding extra amounts of those specific ingredients, since the flavor profile would still not remain constant over the storage period.   However, if one could predict the sorption behavior based on some inherent properties of the solutes and polymers, appropriate optimization of the compensation could be made.   Thus, it was of interest to determine the relative sorption behavior of these different flavor ingredients as a function of structure and functional groups and employ the data to attempt to understand the basis for the differences in sorption.   With this in mind, we determined the effects of permeant structure on sorption in polymers of similar polarity but different chain structure.

The polymers chosen are shown in Figure 1. Low density polyethylene (LDPE) is a branched polyolefin used as a liner for aseptic packaging of juices and for coating paperboard juice containers.   High density polyethylene (HDPE) is relatively unbranched and shows barrier properties superior to those of LDPE.   Isotactic polypropylene (PP) is a polyolefin with methyl groups attached to the backbone in a stereoregular configuration which crystallizes in a helical conformation in contrast to the zig-zag extended chain conformation of polyethylene.   It is used in food packaging multilayer films and containers for liquids as the hydrophobic barrier layer to protect high oxygen barrier layers such as ethylene vinyl alcohol copolymers (EVOH).   The terpene flavor compounds are shown in Figure 2.   They were chosen to provide a range of structures similar to those found in orange juice *(10)* that would also serve to illustrate the effects of structure and functional groups.   They included hydrocarbons (limonene, myrcene and α-pinene) which represent open chain and closed ring structures, two aldehydes (citral a and b), one ketone (carvone) and one alcohol (linalool).

Experimental

Materials. Flavor ingredients were obtained from Aldrich Chemical Co. in the following purities: l-carvone (98%), citral a and b (95%), d-Limonene (97%), dl-Linalool (99%), Myrcene (tech, 95%), dl-α-Pinene (97%).   Their purity level was verified in our laboratory by gas chromatography. They were stored in the dark at 4°C to retard degradation. Solutions were prepared at appropriate concentrations in HPLC grade water and 0.2% (v/v) Tween 80 emulsifier (ICI America) and 2.5% (v/v) HPLC grade ethanol. Composition was verified by HPLC analysis of the solutions before each use.

Polymers were supplied as packaging films (Low Density Polyethylene of 0.916 density and High Density Polyethylene of 0.941 density from Union

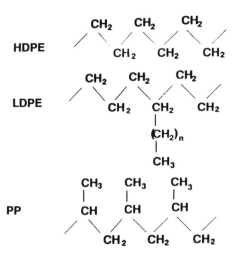

Figure 1. Polymer chain structures of high density polyethylene (HDPE), low density polyethylene (LDPE) and Isotactic polypropylene (PP).

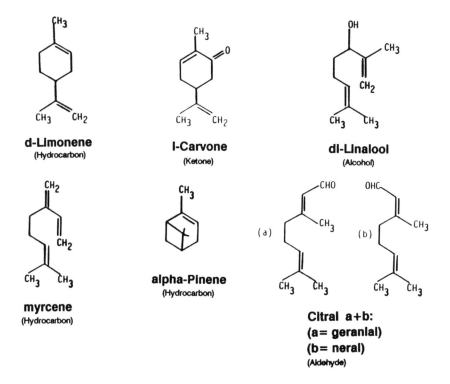

Figure 2. Chemical structures of model compounds.

Carbide Co. and Isotactic homopolymer polypropylenes of densities 0.902 and 0.905 from Shell Chemical Co. They were ground in a Spex freezer mill (Spex Industries, Metuchen, NJ) under liquid nitrogen to increase surface area for sorption and were dried in a vacuum oven for 24 hrs. at $50°C$ before use. They were sieved through U.S. Standard 80-100-150-200 mesh sieves and the 100-150 mesh fraction (106-150 microns) used to provide a uniform surface area.

Analysis of Flavor Compounds. The concentration of the flavor compounds in the model systems was determined by HPLC. A reverse phase analytical column (Varian Micropak C18) in a Varian LC 5000 liquid chromatograph with 4270 integrator was used with an isocratic mixture of water: methanol (25:75). Flow rate was 1 ml per min. at a pressure of 155 atm. UV detector was set at 210 nm which was satisfactory for all of the compounds. Sample injections were made with a $100\mu\ell$ Hamilton syringe into the $10\mu\ell$ loop of the injection valve. Quantitation was by comparison of peak areas with those of known concentrations of the authentic compounds.

Sorption Experiments. Short term (4 hour) experiments were carried out to focus on kinetic behavior in the first minutes after contact with the polymer. Solutions contained 200 ppm of each solute and 3.68 micromoles/liter of $NaNO_3$ as internal standard. Two ml of solution was placed into 13x100 mm glass test tubes containing 40 mg of polymer powder. Two mℓ of solution was also measured into other tubes without polymer as controls. The master solution and the individual sample tubes were all maintained at $23°C$ in a water bath with a test tube shaker. Solutions were separated from polymer at 1, 3, 5, 20, 60, 120 and 240 minutes by sucking through 10 mm glass wool in the tip of a 150 mm glass Pasteur pipet. Ten $\mu\ell$ samples were drawn out for HPLC analysis. Four analyses were taken for each sample and each control for each time period. Sorption vs. time was calculated by comparing the concentrations of the solutions containing polymer with the concentrations of the control tubes and was reported as percentage of initial concentration remaining in solution.

Long term (30 day) experiments were carried out to provide partitioning data. Solutions contained carvone at 200 ppm or limonene at 1000 ppm and $NaNO_3$ internal standard as above. Limonene was initially tested at a concentration close to orange juice, at 200 ppm, but it was found to be almost fully absorbed early in the long time study and led to use of 1000 ppm to permit sufficient residual solute for accurate analysis. Two ml of solution was placed into 2 ml ampules containing 40mg of polymer powder and flame sealed. Control samples without polymer were sealed at the same time. All samples were covered with aluminum foil to protect against light and shaken on a gyratory shaker at 250 rpm at $23°C$. Individual ampules were broken open and analyzed after 1, 5, 10, 15 and 30 days. Four analyses were made on each of two samples and controls for each time period. Sorption was plotted as the difference between the concentrations of the control and experimental tubes.

Statistical Analysis. All sorption data were analyzed for mean values and standard deviations at each time interval and evaluated for significance of differences in sorption of each compound in a given polymer at the 95%

confidence limits using the students' t-distribution for the area between $\pm$ 2.5% for each tail. Specific results are noted in the discussion.

Results and Discussion

Behavior of Mixture of Solutes in Aqueous Solution

An experiment to determine any differences in sorption behavior among the terpenes towards polypropylene was made by placing a mixture of the seven terpene compounds at 200 ppm each in aqueous solution in contact with polypropylene. The method was as described for the short term experiments. Significant differences in behavior were found between the hydrocarbon monoterpenes and the polar oxygenated monoterpenes. An HPLC chromatogram of these compounds before and after exposure to PP for 4 hours is shown in Figure 3 and sorption vs. time is shown in Fig. 4. Sorption began immediately and formed a definite pattern with the three hydrocarbon terpenes being sorbed significantly more than the more polar terpenes. This same pattern had been found in the case of LDPE *(8)* and the final % sorptions are compared in Table I.

TABLE I

Comparative sorptions of monoterpenes by PP and LDPE from 200 ppm of each solute in aqueous solution mixture

Compound	Class	% Loss from Solution	
		PP (4 hours)	LDPE[a] (25 days)
Linalool	Alcohol	7	13
Carvone	ketone	16	10
Citral a	aldehyde	25	11
Citral b	aldehyde	25	11
Myrcene	hydrocarbon	76	40
Pinene	hydrocarbon	89	30
Limonene	hydrocarbon	90	30

[a]Ref 8

The LDPE data are for 25 days while the PP data are for 4 hours, and indicate that PP sorbs more than LDPE. The important point is that the split into two polarity groupings is the same: the hydrocarbons and the oxygenated terpenes. On the basis of these comparisons, further detailed examination of the differences in behavior was made with limonene as representative of the hydrocarbon terpenes and carvone as representative of oxygenated terpenes with similar structures. The behavior towards different types of polymer branching was studied using LDPE, HDPE and stereoregular polypropylene.

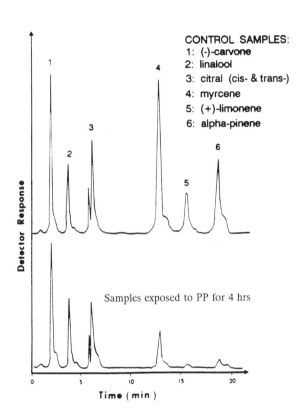

Figure 3. HPLC chromatograms of model compounds before and after 4 hour contact with polypropylene.

Comparative Sorption Behavior of Limonene and Carvone

Short term. The kinetic differences between limonene and carvone towards each of the polymers was clearly seen during the first minutes. In each case, the behavior pattern was the same: an immediate sorption during the first minute followed by a levelling off to a much lower rate. The slope for limonene was much steeper than for carvone and limonene was always sorbed significantly more than the carvone. A typical time course plot is shown in Figure 5 for sorption by PP. At 200 ppm concentrations, limonene was sorbed 92% while carvone averaged 14%. The plots with LDPE and HDPE were quite similar to that of PP in Figure 5. The four hour losses are summarized in Table II. It appeared that the three differently branched polyolefins did not affect limonene sorption but did affect carvone.

Table II
Comparative sorptions of limonene and carvone by LDPE, HDPE and PP
from 200 ppm of each solute in aqueous solution at 23°C in 4 hours

Polymer	% Loss from Solution	
	Limonene	Carvone
LDPE	94	32
HDPE	94	23
PP	92	14

Long term. The next set of experiments consisted of 30 day sorptions in each of the polymers to determine comparative partition coefficients. The results are shown in Figures 6 and 7. Limonene was sorbed more than carvone in every case. Figure 6 shows that limonene was sorbed more by LDPE than by HDPE (61 vs 43%) and that PP sorbed limonene only slightly more than LDPE (64 and 69% vs 61%) but more than HDPE (43%). Thus, the order of affinity of limonene sorption was PP = LDPE > HDPE. The coefficient of variation for these experiments was about $\pm$ 2% and demonstrated that the differences were significant at the 95% Confidence Limits.

The carvone sorption plotted in Figure 7 shows it too was sorbed more by LDPE than by HDPE and the difference was significant (95% Confidence Limits of 33 to 37 vs. 22 to 26). The data for PP show less sorption than LDPE and straddle the data for HDPE (95% Confidence Limits for PP_1 = 19-21% and PP_2 = 29-30%). By these measurements, the order of affinity of carvone is LDPE > HDPE = PP.

Partition Coefficients. The long term sorption data were employed to estimate partition coefficients (Kp) for limonene and carvone between the individual polymers and the aqueous solutions. They were calculated from the expression:

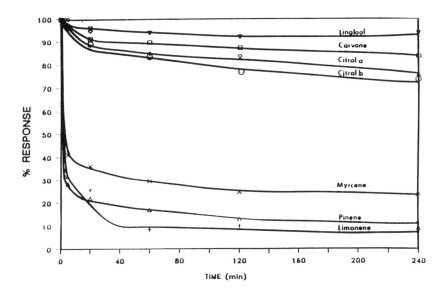

Figure 4. Sorption of mixture of seven model compounds at 200 ppm into polypropylene at 23°C.

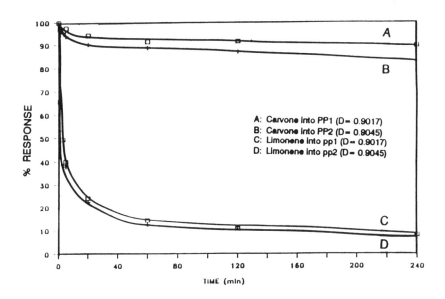

Figure 5. Sorption of carvone and limonene at 200 ppm into polypropylene at 23°C.

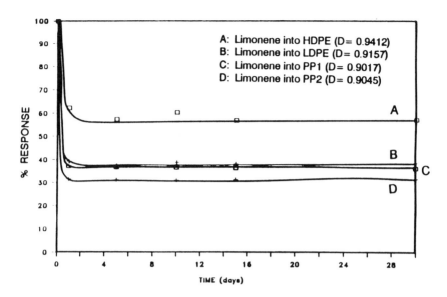

Figure 6. Sorption of limonene at 1000 ppm at 23°C into high density
polyethylene (HDPE), low density polyethylene (LDPE) and
polypropylene (PP1 and PP2).

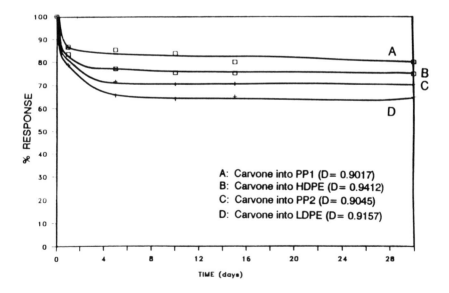

Figure 7. Sorption of carvone at 200 ppm at 23°C into high density
polyethylene (HDPE), low density polyethylene (LDPE) and
polypropylene (PP1 and PP2).

$$Kp = \frac{\text{Concentration of solute in polymer}}{\text{Concentration of solute residual in solution}}$$

The concentrations were determined after 30 days using the values for the 2 mL aqueous samples in contact with 40 mg of polymer. The Nernst distribution coefficient on which this calculation is based actually calls for absolute equilibrium concentrations, however, since concentration values were little changed after the 5th day, this expression was satisfactory for comparative estimates. The Kp's are tabulated in Table III. For each of the polymers, limonene has a significantly higher Kp than does carvone. The table also lists the changes in Gibbs free energy (ΔG) calculated from these Kp values with the expression:

$$\Delta G = -RT\ln Kp$$
where R = 1.987 cal/g mole $^\circ K$
$\quad\quad T$ = $296^\circ K$

Table III

Comparative partition coefficients (Kp) and Gibbs free energy changes (ΔG) for sorption of limonene and carvone by polymers from aqueous solutions at 23°C

Polymer	Limonene		Carvone	
	Kp	ΔG^a	Kp	ΔG^a
LDPE	79	-2.6	27	-1.9
PP	98	-2.7	17	-1.6
HDPE	35	-2.1	16	-1.6

[a] = Kcal/g mol $^\circ K$

The Gibbs relation is based on Kp at equilibrium and we have assumed that these Kps are at conditions close enough to equilibrium to be used for purposes of comparison. The resultant error in ΔG is regarded as small. The value of the partial molar free energy change (ΔG) gives an estimate of the strength of the tendency for sorption; the more negative the value of ΔG, the stronger the sorption tendency. The table shows that in each polymer, the interaction with limonene is stronger than with carvone. These results would indicate that solute polarity was the predominant controlling factor and that polymers with similar low polarities had a lesser effect on sorption.

Theoretical Considerations

The theoretical basis for the sorption of the flavor ingredient from the solution into the polymer involves both partitioning and diffusion. It is worthwhile to examine the relative balance of the two factors, partitioning and diffusion, and their application to the occurrences under the conditions of our experiments. The mechanism of sorption consists of adsorption onto the solid surface followed by dissolution into the polymer and diffusion away from the surface under the driving force of a concentration gradient until equilibrium is established. The mathematics have been worked out in detail and can be found in several reference books *(11, 12, 13)*. Those references describe a number of theoretical approaches and refinements for considering the balancing of these factors. We shall select among them and attempt to apply it to our data. The concept is the interplay among cohesive forces in the system. This involves two considerations: 1) the partitioning step involves adsorption onto the polymer surface and mixing into the polymer matrix. This depends on the relative forces of attraction for the solute between the solution and the polymer. These forces are governed by the thermodynamics of the system and are affected by structural and polarity effects. These factors can be considered on the basis of cohesive energy density forces or its square root, the solubility parameter of the Hildebrand equation: $\Delta H_m = v_1 (\delta_1 - \delta_2)^2 \phi^2_2$ where ΔH_m is the partial molar heat of mixing, v_1 is the partial molar volume of the penetrant, ϕ_2 is the volume fraction of the polymer in the mixture and δ_1 and δ_2 are the square roots of the cohesive energy density (CED) of the penetrant and the polymer, respectively *(14)*. The differences in the CED values of polymer-penetrant combinations in a given sorption system will affect the size of the heat of mixing and a permeant with a larger CED will have a larger heat of mixing and less negative ΔG than one with a smaller CED under the same circumstances. 2) the diffusion step depends on the mobility of the solute which can depend on its size and its interaction with the polymer to plasticize and relax its chain segments for cooperative hole formation. Size can be affected by CED if it favors clustering of the solute molecules by cohering to one another. For example, if two compounds have a similar molecular size but one of them has a stronger tendency to cohere and cluster, then the clustered compound can act as a larger solute *(11)*. Further, its tendency to interact with and plasticize the polymer would also be reduced. These differences can have a strong effect on lowering the diffusion rate.

Mechanistic Implications

These factors can be applied to our observed data and be examined for relevence to providing a reasonable interpretation of the behavior of the chosen solutes and polymers. The main observations can be summarized as follows:

1.   In a mixture of seven terpenes of different molecular structures and functional groups, the hydrocarbon solutes (limonene, myrcene, pinene) were sorbed significantly more than the more polar alcohol, aldehyde, ketone solutes (linalool, citral a and b, carvone) by polyolefins of different branched structure but similar polarities.

2. Comparison of the hydrocarbon limonene and the ketone carvone of similar structure but different polarity in sorption by LDPE, HDPE and PP always showed that the initial sorption of both compounds was almost immediate but the limonene was sorbed at a faster rate and to a greater extent.

3. In long-term experiments, the order of affinity for limonene, as judged by K_p was PP = LDPE > HDPE while for carvone it was LDPE > HDPE = PP.

These observations, along with kinetic and thermodynamic data reported for limonene and carvone sorption with LDPE *(15)* can be interpreted on the basis of balancing of cohesive forces.

Table IV shows a compilation of cohesive forces for the terpenes and polymers used in these experiments. They were calculated here on a self-consistent basis from tables of structural group molar cohesive energy factors for liquids *(16)* and polymers *(17, 18, 19)*. The method of calculation consists of summing the factors for molar vaporization energy (U) and for molar volume (V) for each of the molecular structural elements. These sums are then employed in the equations for CED = (-U/V) and for δ = $(-U/V)^{1/2}$, where CED is the cohesive energy density in cal cm^{-3} and δ is the Hildebrand solubility parameter in cal$^{1/2}$ cm$^{-3/2}$. The latter value is also listed in SI units of MPa$^{1/2}$ by conversion of cal$^{1/2}$ cm$^{-3/2}$ to J$^{1/2}$ cm$^{-3/2}$ which is equivalent to MPa$^{1/2}$.

Table IV
Calculated values of cohesive energy densities and Hildebrand solubility parameters for selected terpenes and polymers at 25°C

Compound	$\dfrac{CED}{Cal\ Cm^{-3}}$	$Cal^{1/2}Cm^{-3/2}$	δ $MP_a^{1/2}$
Polypropylene	66	8.10	16.6
Polyethylene	66	8.10	16.6
Myrcene	62.7	7.92	16.2
Limonene	67.4	8.21	16.8
α-Pinene	70.3	8.38	17.1
Linalool	84.3	9.18	18.8
Carvone	89.5	9.46	19.4
Citral a	102.5	10.12	20.7
Citral b	102.5	10.12	20.7

This table can be used to compare the cohesive forces acting on the solutes and polymers of our experiments.

For case 1 of the summarized observations, it can be seen that the hydrocarbons have cohesive forces close to those of the polymers (CED in the range of 63 to 70 cal cm^{-3}) while those of the oxygenated compounds are higher, (CED of 84-103). As an example of the effect of these differences, after converting CED to δ, calculation of comparative heats of mixing from the Hildebrand equation $\Delta H_m = v_1(\delta_1 - \delta_2)^2 \phi_2^2$ gives for limonene (16.8-16.6 MPa$^{1/2}$)^2 = 0.04 MPa$(v_1\phi_2^2)$ while carvone would be (19.4-16.6 MPa$^{1/2}$)^2 = 7.84 MPa $(v_1 \phi_2^2)$. The ratio of the differences would be 7.84/.04 = 196 times higher heat of mixing for the carvone compared to limonene under the same set of conditions. Similar differences would be found for the other hydrocarbons and oxygenated compounds and would illustrate the magnitude of the differences in cohesive forces arising from the relative structures and functional groups of the solutes to yield the two groupings observed in the experiments. They are reflected in the ΔG values in Table III.

For case 2 involving immediate sorptions of both limonene and carvone followed by different long term sorptions, the same reasoning would apply to the partitioning and diffusion involved in sorption. In the initial sorption phase (adsorption), carvone would be slower than limonene because its higher cohesive energy would retard the rate of interacting with the attractive centers of the polymer. That is, penetrant-penetrant attractions would be stronger than penetrant-polymer attractions and retard solubility in the polymer. That is shown in the relative CED values. This would also result in retardation of carvone sorption during the second phase (absorption), because of its greater cohesion compared to limonene. Added to this would be the retarding effect of any of carvone clustering to present a larger structure hindering mobility within the polymer. In general, the mobility is considered to be a factor of $(\sigma)^n$, where σ is the molar cross section *(11)*, and n varies between 1 to 2 depending on shape and orientation. Thus the larger cohesive forces in carvone would lead to a significantly lower diffusion rate than that of limonene. These differences in behavior were observed with all three of the polymers and reflect differences in the sorption mechanisms for the two solutes. Based on kinetic and thermodynamic data, it has been shown previously *(8, 15)*, that sorption of limonene by LDPE was biphasic, showing both adsorption and absorption, while carvone exhibited predominantly adsorption in the same system. The same behavioral differences may now be involved with all three of the polyolefins, even though their branching structures differ.

The third part of the observed data shows differences in the order of affinity between the solutes and the polymers. This may be a result of the differences in cohesive forces of the solutes that would affect differences in diffusion in the free volumes in the polymers as a result of their structures. The order of affinity of limonene was PP = LDPE > HDPE. Because of similarity of CED of limonene and these polymers, much penetrant-polymer interaction could occur to yield plasticization and polymer-segment relaxation. Differences in the packing density of the unbranched HDPE compared to the LDPE with its short chain branches and to the PP with its methyl groups on alternate carbons could result in a higher susceptibility to plasticization for the LDPE and PP. This would increase their free volume and could account for the order of affinity observed.

In the case of carvone, the cohesive forces would result in a lower degree of penetrant-polymer interactions and less plasticization of the polymers. In that case, its penetration would have to depend on the already existing differences in free volumes among the polymers. This would place the polymers in the order LDPE, with its chain segments held apart by its short chain branches >HDPE = PP where regularity of the chain structures permit better chain packing. To the extent that these structural relationships were not changed by plasticization by carvone, the order would be the same as that actually observed in this study.

In conclusion, data from the experiments have shown that structure and functional groups of the solutes exerted a strong influence on sorption behavior towards polymers of similar polarities. The hydrocarbon solutes were sorbed faster and to a greater extent than the oxygenated solutes. Polymer structural differences appeared to influence the degree of sorption resulting in two different modes of sorption behavior and orders of solute-polymer affinities. These observed behaviors could be interpreted on the basis of differences of cohesive forces that can affect the course of the partitioning and diffusion steps. This interpretation could provide a basis by which to predict the comparative compatibility of food solutes with polymeric packaging materials.

Literature Cited

1. Halek, G. In *Food and Packaging Interactions*; American Chemical Society Symposium Series 365; *1988.*, p. 195-202. J. Hotchkiss, Ed.
2. Dürr, P., Schobinger, U., Waldvogel, R. *Alimenta. 1981*, *20*:91.
3. Dürr, P. and Schobinger, U. In *Flavour '81*; Scheier, P. Ed., Walter de Gruyter and Co.,: Berlin, New York. *1981*, p. 179-193.
4. Marshall, M., Adams, J. and Williams, J. *Proceedings Aseptipak '85*, *1985*, Princeton, NJ.
5. Mannheim, C., Miltz, J. and Letzler, A. *J. Food Science*, *52*:737-740.
6. Kwapong, O. and Hotchkiss, J. *J. Food Science*, *52*:761-763 and 785.
7. Hirose, K., Harte, B., Giacin, J., Miltz, T. and Stine, C. In *Food and Packaging Interactions*, American Chemical Society Symposium Series 365, *1988*, p. 28-41.
8. Halek, G. and Meyers, M. *Packaging Technology and Science, 2*, 141-146 *1989*.
9. Imai, T., Harte, B. and Giacin, J. *J. Food Sci. 55*:1, 158-161.
10. Nisperos-Carriedo, M. and Shaw, P. *Food Technology - April 1990*, p. 134-138.
11. Rogers, C. In *Polymer Permeability*; Comyn, J., Ed., Elsevier; New York, *1985*; Chapter 2.
12. Rogers, C. In *Physics and Chemistry of the Organic Solid State*; Fox, D., Labes, M., Weissberger, A., Eds.; Interscience: New York, *1965*; *Vol. II*, Chap. 6.
13. Crank, J. *The Mathematics of Diffusion*; 2nd Ed., Clarendon: Oxford, 1975.

14. Hildebrand, J. and Scott, R. *The Solubility of Nonelectrolytes*, 3rd Ed., Reinhold, New York, *1950*.
15. Meyers, M. Ph.D. Thesis, Rutgers University, New Brunswick, NJ, *1987*.
16. Barton, A., *CRC Handbook of Solubility Parameters and Other Cohesion Parameters*; CRC Press: Boca Raton, FL, *1983*.
17. Barton, A. *CRC Handbook of Polymer-Liquid Interaction Parameters and Solubility Parameters*; CRC Press: Boca Raton, FL, *1990*.
18. Fedors, R. F. *J. Polymer Science, C 26*, 189. *1969*.
19. Van Krevelen, D. W. and Hoftyzer, P. *Properties of Polymers*; Elsevier: Amsterdam, *1976*.

RECEIVED March 1, 1991

Chapter 19

Scalping from a Paste into a Series of Polyolefins

Absence of Correlation between Quantity Sorbed and Polymer Crystallinity

Patrick Brant[1], Dirk Michiels[1], Bert Gregory[2], Keith Laird[1], and Roger Day[1]

[1]Exxon Chemical Company, 5200 Bayway Drive, Baytown, TX 77522
[2]Exxon Chemical Company, Nieuwe Nijverheidslaan 2, B–1920, Machelen, Belgium

A general method for evaluating scalping from pastes by polymers has been applied to scalping from a mint-flavored toothpaste. The quantities of six key volatile flavorants - d-limonene, 1,8-cineole, menthone, menthol, (-)carvone, and cinnamaldehyde-scalped from the toothpaste by a given polymer are determined by multiple extraction head space gas chromatography (MEHSGC). Relative extents of scalping by the polymer films are also measured gravimetrically. The scalping characteristics of the following polymers are compared: HDPE, MDPE, VLDPE, LLDPE, an ethylene vinylacetate copolymer (EVA), an ethyleneacrylic acid copolymer (ESCOR 5100), and two zinc ionomers (IOTEK 4000, EX 906). The data obtained in this study indicate that the relative extents of scalping of the eight different polymer films studied increase in the following order:

MDPE < LLDPE ~VLDPE < HDPE < ESCOR 5100 < EVA < IOTEK 4000 < EX 906

Thus, the extent of scalping is not at all related to the crystallinity of the polymer. It is, on the other hand, very much related to the presence of oxygen containing comonomer and to ionomerization. Additionally, scalping is shown to be very inhomogeneous (quantities scalped are not in proportion to quantities found in toothpaste).

Polymers offer many advantages for packaging food and personal care products. These are by now well known. They can include light weight, microwavability, and low cost, among others. However, they can also present two disadvantages. These are also well documented. As illustrated in Figure 1, they include "flavor scalping", or the transport of important flavor/aroma molecules from the food into the package, and "off-taste", or the transport of molecules from the polymer to the food or health care product.

0097–6156/91/0473–0227$07.00/0
© 1991 American Chemical Society

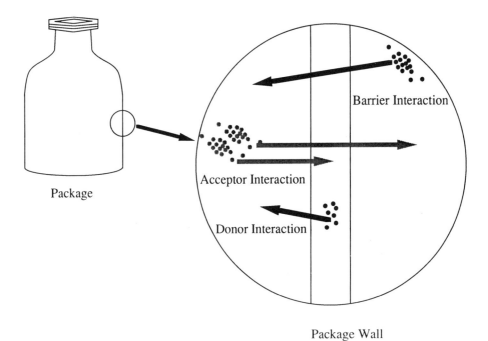

Figure 1.   Donor (off-taste), acceptor ("scalping"), and barrier interactions in a packaged product.

In this work, we examine scalping of flavor/aroma molecules from toothpaste. It might be surprising to most, but flavor scalping is substantial for toothpaste packaged in flexible tubes. In a commonly used flexible tube construction, the toothpaste contact layer can be polyethylene or a related polyolefin (Figure 2). Aromatic molecules in the toothpaste are sorbed by the polymer contact layer. One strategy for counteracting the effects of scalping is incorporation of 5-10% excess of the flavor "cocktail" in the paste. It is estimated that this counteraction costs between 10 and 18M$ each year in the United States alone, so there is clearly some incentive for reducing scalping. This study was undertaken to determine what sort of changes in scalping could be effected by changing polymer structure.

Toothpastes are complex mixtures. Like other pastes, they are not readily removed from the polymer contact layer. This can cause difficulties in most experiments one might try to measure scalping extents by different polymers. We are not aware of any previous published work regarding scalping from pastes. For this initial foray, the composition of a "representative" toothpaste had to be determined and a method had to be developed for studying scalping from pastes. Toothpaste composition was determined by a combination of purge-and-trap GC-mass spectrometry and multiple extraction head space GC (MEHSGC). The scalping method used employs a sacrificial polymer contact layer (Figure 3). This method had been used with success in previous measurements of solubilities of polymer additives in various polymer films. It is a modification of a method reported in the literature.[1,2] The quantities of each of six key aroma/flavor molecules sorbed into each polymer for a given contact time were determined by multiple extraction head space gas chromatography[3-6]. In addition, the weight gains of polymer films contacting the toothpaste were determined gravimetrically. At the end of each sorption experiment, a weighed portion of each film was held in vacuo at 55-70°C to constant weight. The weight loss was used as a second measure of total quantity volatile aroma molecules sorbed.

Experimental

Polymer Films. The polymer films used in this study (film thicknesses, in mil, in parenthesis) are listed in Figure 4. These polymers span a wide range of crystallinity (23-72%) and comonomer type, or polymer polarity.

Sorption Experiments. The scalping test is carried out as shown in Figure 3. Inside a nitrogen purged glove bag, three to five layers of each film are stacked (air bubbles pressed from between layers) over the mouth of a one liter jar into which has been placed 100g of a commercially available mint toothpaste. Aluminum foil is then fitted over the films and a canning jar lid screwed on to seal the system. Each jar is turned upside down so that the paste contacts the innermost film. All jars are left inside the nitrogen purged bag until the films are removed for analysis.

Measurements. Volatiles contents of the films were determined by multiple extraction head space gas chromatography (MEHSGC; Perkin-Elmer HS100 autosampler coupled to Sigma 2000 gas chromatograph 60m DB-5, 0.32mm ID column, flame ionization detector; sample vial equilibrated at 120°C). Volatiles in mint-flavored toothpaste were also analyzed by a purge-and-trap method using a Tekmar Model 4200 coupled to a Hewlett-Packard 5890A gas chromatograph (also operated with 60m DB-5 0.32mm ID column and FID). A splitter was used so gas

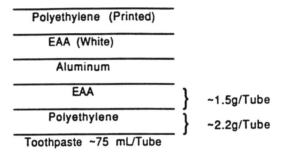

- Apparently, about 5-10% of flavor and aroma components in toothpaste are scalped by the polymer contact layer.

- A common strategy used to counteract the effects of scalping is incorporation of 5-10% excess of flavor "cocktail" in formulation.

Figure 2.   Example of toothpaste tube construction.

PASTES

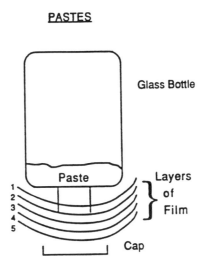

Throw away layer 1 (contact layer).
Use layers 2-5 for analysis

The toothpaste is maintained in contact with the layers of polymer until the concentrations of volatile flavor molecules are approximately constant through all the layers of film. In practice, for a stack of four films, each 0.05mm thick, it takes about 20 days to approach equilibrium.

Figure 3.   Schematic of "sacrificial" layer sorption experiment.

components could be analyzed with both a flame ionization detector and a Hewlett-Packard 5790B mass selective detector. Flavor compounds identified by PATGC-mass spectrometry in the toothpaste were corroborated by comparing retention times of standards with those found in the HSGC of the toothpaste.

In one series of sorption experiments, portions of polymer films were maintained under vacuum for over 24 hours prior to use to obtain a stable weight for the samples. After the toothpaste sorption step, a piece of each film was weighed. These weighed portions of films were then placed in a vacuum oven at 55-70°C until each reached constant weight (roughly 24 hours). The weight loss here was taken as a further measure of the quantity sorbed of volatile flavor/aroma molecules.

Results and Discussion

Identification of Volatile Components in Toothpaste. A HSGC of the toothpaste employed in this study is shown in Figure 5. The volatiles profile of the toothpaste was also examined by PAT-GC-MS. A total ion chromatogram (TIC) was recorded. Molecular assignments based on mass spectral fragmentation patterns and matching retention times with those for standards were made. Key volatile molecules identified in the toothpaste both by mass spectral fragmentation pattern and by retention time match with standard compounds include the following (concentrations, in wppm, in parentheses): d-limonene (386), 1,8-cineole (279), menthol (863), menthone (2185), (-)carvone (1428), and cinnamaldehyde (278). Structures of these six molecules are shown in Figure 6. Some of the more important properties of these molecules are listed in Table I. Calculated solubility parameters for these molecules and for some of the polymers used are listed in Table II.

Multiple Extraction Head Space GC Data. Multiple extraction HSGC experiments were carried out at 120°C. Ln (area) versus extraction number plots are shown for some of these samples in Figures 7 and 8. These plots indicate that the desorption behavior of the molecules from the films follows Henry's Law. In order to obtain quantitative information, the MEHSGC data must obey natural decay kinetics. In these cases, all data do. Linear leasts squares fits of the data typically give linear correlation coefficients of greater than 0.99. The slopes of these lines are (for samples weighing 0.05-0.06 g) typically in the range 0.08 to 0.24. Concentrations of six key volatile molecules were calculated from the data. Data collected for all three or four film layers clearly indicate that an apparent or pseudo equilibrium through the different layers of film is reached at contact times of 20-30 days (at ambient temperature).

The accuracy of the MEHSGC measurements should be ±5-10%. We tested the accuracy by making up a standard aqueous emulsion with known quantities of menthol, menthone, and carvone in it. The emulsion was analyzed twice, once right after it was made and again after "aging" five days. The quantities of aroma molecules found in the emulsion were equal to those placed in the emulsion within ±5% (Table III). So these "calibration" experiments illustrate, as have other similar calibrations, that MEHSGC can be quite accurate.

The quantities of each of the six important flavor/aroma molecules were measured in each film and in the mint toothpaste by MEHSGC at 120°C for contact times of 20-45 days and 70-130 days.

For a simple system, if equilibrium were reached, we might expect the concentrations of the six volatile molecules to be equal for the two sampling times of 20-45 days and 70-130 days. In general, as shown in Table IV, the concentrations

Polymer	% Crystallinity	Film Thickness, mil
HDPE	72	2.0
MDPE	46	3.1
LLDPE	39	1.5
Escor 5100 (EAA)	41	1.8
18% EVA	37	1.2
IOTEK 4000 (Zn ionomer)	~41	2.0
EX906 (An ionomer)	~30	2.0
VLDPE	25	2.0

Figure 4.   List of polymer films tested, their crystallinities, and thicknesses.

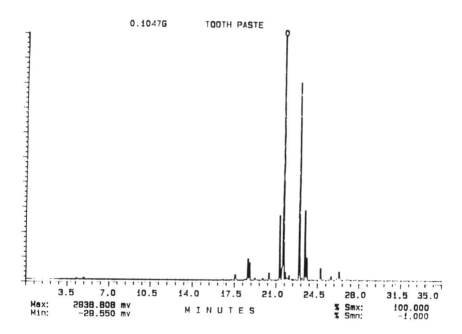

Figure 5.   HSGC of mint toothpaste.

CONCENTRATIONS IN TOOTHPASTE, WPPM, GIVEN IN PARENTHESES

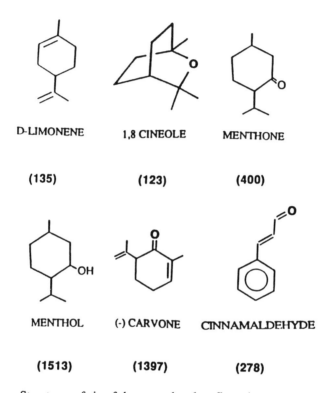

Figure 6.   Structures of six of the most abundant flavor/aroma molecules in mint toothpaste.

Table I. Some Properties of Key Flavor Molecules in Mint Toothpaste

Compound	MW	Refractive Index @ 20°C	H_{VAP}[a]	Density, g/cm^3 @ 20°C	B.P., °C
Menthone	154.3	1.4505	4611	.8963	209.6
Menthol	156.3	1.461	4760	.904	216.4
Carvone	150.2	1.4999	5082	.9608	231
Limonene	136.3	1.4730	3738	.8411	178
Linalool	154.3	1.4636	9378	.8700	199
1,8 Cineole	154.3	1.4586	3880	.9267	176.4
Cinnamaldehyde	132.2	1.6195	5565	1.0497	253

[a] Estimated using Trouton's Rule: $\Delta S_{VAP} \cong 21$ cal/deg-mol $= \Delta H_{VAP}/T_{BP}$
For polar molecules: $\Delta S_{VAP} \sim 22$ cal/deg-mol

Table II. Calculated Solubility Parameters for Important Volatile Compounds in Toothpaste Compared with Solubility Parameters for Some Polymers[a]

Compound	Calculated Solubility Parameters $(MPa)^{1/2}$			
	δ_d	δ_p	δ_h	δ
Carvone	16.2	4.9	4.5	17.5
Cinnamaldehyde	19.94	6.35	5.98	21.76
1,8 Cinneol	14.30	2.4	4.25	15.11
D-Limonene	15.25	0	0	15.25
Menthol	18.34	2.9	10.76	21.46
Menthone	15.09	4.5	3.4	16.10
Myrcene	15.85	0	0	15.85
Pinene	15.70	0	0	15.70
Polyethylene[b]	17.0	0	0	17.0
Polypropylene[b]	17.3	0	0	17.3
EVA, 18%[b]	16.8	2.0	~2	17.7

[a] Calculated using group molar dispersion, polar, and hydrogen bonding contributions F_d, F_p, and U_h, respectively. Taken from "CRC Handbook of Solubility Parameters and Other Cohesion Parameters", Albu F. Barton, author, CRC Press, Inc. Boca Raton, FL.

[b] Estimated or taken from data in "CRC Handbook of Solubility Parameters and Other Cohesion Parameters."

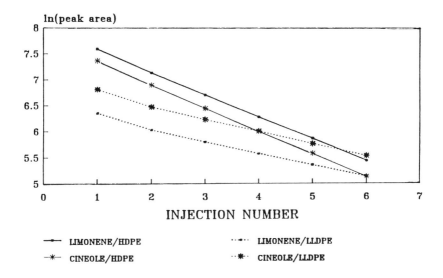

Figure 7. MEHSGC at 120°C; Ln (area) versus injection number for d-limonene and cineole sorbed into HDPE and LLDPE.

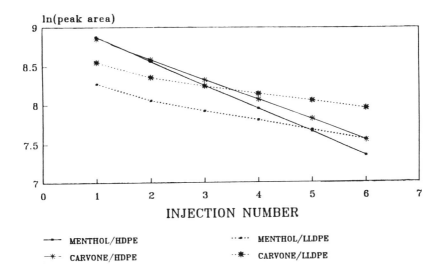

Figure 8. MEHSGC at 120°C; Ln (area) versus injection number for menthol and carvone sorbed into HDPE and LLDPE.

Table III. Mass Balance Experiment Flavor Molecules in Aqueous Emulsion[a]

Compound	Amount Placed by Emulsion, by weight[b]	Amount Found in Emulsion, by MEHSGC	
		Start	After 5 days
Menthol (R.F. 60.9)[c]	2005	2048	2126
Menthone (R.F. 60.8)	1007	948	1050
Carvone (R.F. 54.9)	1539	1639	1636

a   Made using sodium octyl sulfate/sodium lauryl sulfate mixture
b   ppm
c   response factor

Table IV. Sorption of Volatile Components from Mint Toothpaste into Polymer Films

Sample	Concentration of Volatile Component, wppm [a]						"Total" Volatiles[c]
	D-Limonene 1	1,8 Cineole 2	Menthone 3	Menthol 4	(-)Carvone 5	Cinnamaldehyde 6	
HDPE [2.0][b]							
Layer #3 (22)	871	708	3497	3898	5164	2237	15,375
(100)	126	258	2123	4850	5335	2027	14,719
MDPE [3.1]							
Layer #3 (29)	173	532	2284	2440	4445	1402	11,296
(80)	23.9	119	584	1539	2701	631	5,598
LLDPE [1.5]							
Layer #3 (22)	327	501	3489	3475	6663	3108	17,563
(130)	64.2	136.4	1332	2939	4245	1685	10,402
VLDPE [2.0]							
Layer #2 (23)	84	352	2105	2400	4791	1016	10,748
Layer #3 (100)	104	253	1686	2997	5267	1613	11,920
EVA [1.2]							
Layer #3 (23)	65	122	1056	4186	5746	1675	12,850
(100)	619.1	347	2223	11706	13814	2111	30,820
EX906 [2.0]							
Layer #2 (31)	470	1730	4065	13736	12843	1873	34,767
(70)	145	1162	3782	32677	20467	2653	60,886
ESCOR 5100 [1.8]							
Layer #2 (35)	360	1269	3422	12,558	11,095	1682	30,386
(70)	39	336	1526	14,739	10,668	1416	28,724
IOTEK 4000 [2.0]							
Layer (45)	66	209	2380	28,920	15,450	3848	50,873
Mint Toothpaste	135	123	400	1,513	1,397	278	3,846

a   Determined by MEHSGC at 120°C.
b   Values in brackets are film thicknesses, in mil. Values in parentheses are days in contact with toothpaste.
c   Sum of concentrations of molecules 1-6.

are not equal. Again, in general, one observes a depletion of the lower boiling molecules - limonene, 1,8 cineole, and menthone, but most especially limonene - in the films at long contact times compared with short contact times. These depletions are perhaps best illustrated in the "star" plots shown in Figures 9-12. In part, the depletions might be due to the experimental design. The polymer film/mint toothpaste system is not rigorously closed. Small amounts of excess polymer film are in contact with the rest of the world, so volatile molecules transported to the film edges have an opportunity to evaporate out of the system. This pathway would favor depletion of the smaller, lower boiling molecules, particularly those not engaged in any specific interactions with the polymer matrix. On the other hand, some of the other depletions measured involving menthone and cinnamaldehyde are not so readily explained by evaporation out of the system. Furthermore, mass balance measurements (see Table IV) suggest that fractions of some molecule concentrations are sorbed back into the toothpaste at long contact times.

In contrast to some of the decreases in concentration with time noted above, significant or very large increases in concentration of some flavor/aroma molecules in the polymer films are also observed. Some particularly large increases are observed for menthol and carvone. Like the decreases, these increases are perhaps more readily conveyed in the star plots in Figures 9-12. In most cases then it appears that the sorption process has not in fact reached equilibrium over the first 20-30 days contact with polymer films. Equilibrium is likely more nearly approached at the long contact times of 70-130 days.

Even though these systems may not have reached equilibrium, the storage times likely approach or equal that expected for a tube of toothpaste and the data do contain important trends or, surprisingly, lack a trend. For example, it is well known that for simple systems (for example, employing a single, non-interacting sorbant instead of a food or personal care product) the quantity of a molecule sorbed is inversely related to the crystallinity (density) of the polymer.[7,8] In this study, such is not the case. We plot the quantity sorbed of flavor/aroma molecules (sum of concentrations of the six key volatiles, as measured by MEHSGC) versus polymer crystallinity[9] in Figure 13. It is clear and very surprising, at least among the polyethylene films, that there is no correlation. This plot shows that factors other than crystallinity are important in the relative quantities of molecules sorbed.

One very gross factor is clearly the polarity (solubility parameter) of the polymer. The oxygenated polymers sorb two to eight times more of the six key flavor/aroma molecules than do the polyethylenes. Along with the summary in Table IV, the relationship is again perhaps more efficiently presented in a star plot (Figure 14). Among the oxygenated polymers, the zinc ionomers sorb much more of the aroma/flavor molecules than the other oxygenates.

Polymer polarity also biases which molecules are more extensively sorbed by the polymer matrix. For example, among the copolymers - ESCOR 5100, EVA, and EX906, and IOTEK 4000 zinc ionomers - we find that a molecule such as menthol with a significant hydrogen bonding component is sorbed in relatively greater quantities compared with menthone, than it is among the polyethylenes. As summarized in Table V, the [menthol/menthone] ratio among the oxygenated polymers is 3.4 to 12.1 whereas among the polyethylenes, it is 1.0 to 2.6. A similar dichotomy is found for the ratio of [carvone/cinnamaldehyde] where among the polyethylenes, the ratio is 2.1 to 4.7, while among the oxygenated polymers, it is 3.4 to 7.7. Presumably, the relatively greater sorption of menthol compared to menthone among the oxygenated polymers arises from hydrogen bonding interactions between the pendant - C(O)O- groups in the polymers and the -OH group on menthol. However, infrared spectra recorded of films before and after

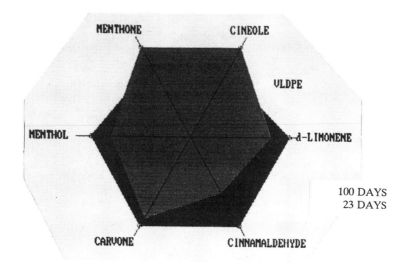

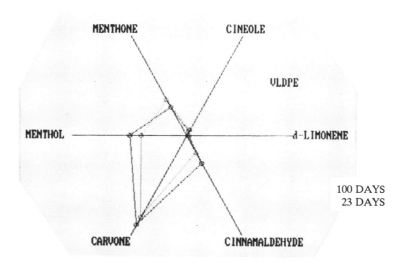

Figure 9. STAR plot comparisons of six key volatiles in VLDPE after short and long contact times with toothpaste. In top plot, the star axes are normalized; in the bottom plot, they are all the same scale.

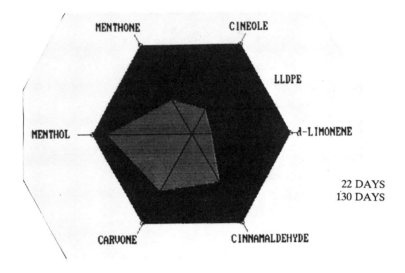

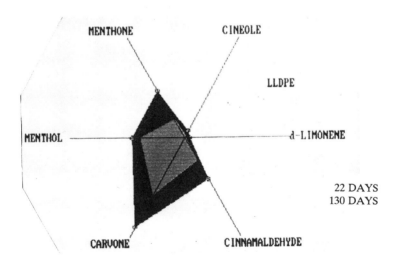

Figure 10.   Same as Figure 9, except plot is for LLDPE.

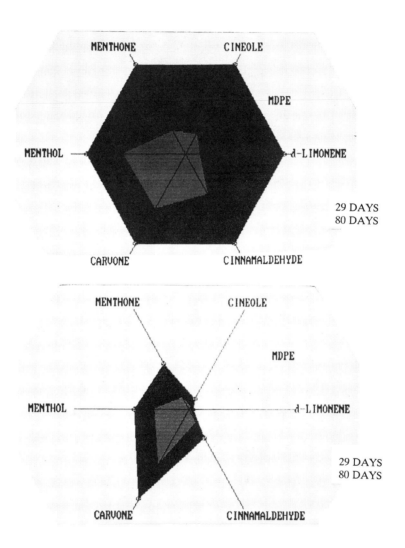

Figure 11.   Same as Figure 9, except plot is for MDPE.

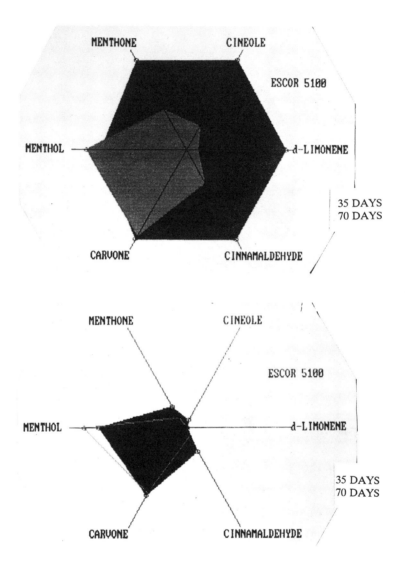

Figure 12.   Same as Figure 9, except plot is for ESCOR 5100 ethylene-acrylic acid copolymer.

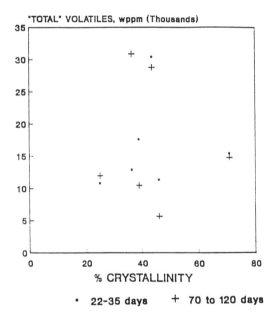

Figure 13.   Plot of sum of quantity sorbed for six key flavor/aroma molecules in polymers films, measured at both short and long contact times, versus polymer crystallinity.

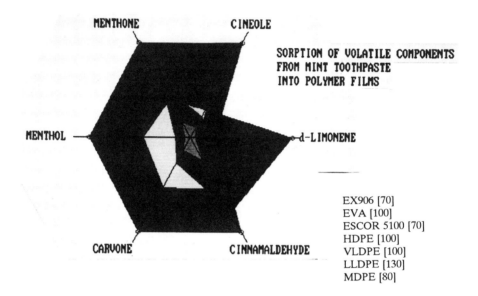

Figure 14.   STAR plot comparison of quantities of six key volatile components sorbed as a function of the polymer contacted with toothpaste.

Table V.   Effect of Polymer Composition on Sorption

Polymer Class	[Menthol/Menthone]	[Carvone/Cinnamaldehyde]
Polyethylenes	1.0 - 2.6	2.1 - 4.7
Oxygenated Polymers	3.4 - 12.1	3.4 - 7.7

sorption of these components do not reveal any detectable shift in the CO stretch (18% EVA; ν_{CO} before 1738.5 cm^{-1}, ν_{CO} after 1738.3 cm^{-1}). One of the most obvious aspects of the data is that by and large the polymers sorb quantities of aroma/flavor molecules which are typically 2 to 15 times greater than their respective concentrations in the toothpaste. In these experiments, one exposes 1 to 2 g. of polymer to 100 g of toothpaste (about the ratio of polymer/toothpaste in a typical tube), so the HSGC data indicate that we are sorbing roughly 2-30% of a given flavor/aroma component from the paste into the polymer. The quantities sorbed are not uniform and are very much dependent on the polymer contacted with the paste.

Gravimetric Data. As part of the series of sorption MEHSGC experiments performed at 120°C, steps were taken to independently measure gravimetrically net weight change of the polymer films - [(weight after sorption)-(weight before sorption)]. All the polymer films were first degassed at room temperature for a period of 2-3 days. Sorption to apparent equilibration was performed. When experiments were stopped (45-100+ days), a portion of each film was weighed (prior to MEHSGC experiments). Then each weighed portion of polymer film which had been in "contact" with the mint toothpaste for 45-100+ days was placed in a vacuum oven for 24 hours or more at 55-70°C. The weight losses were monitored periodically until weight loss between measurements was not detectable.

Excellent agreement between MEHSGC and this gravimetric method was observed. A plot of desorption weight loss (to constant weight) versus the sum of concentrations for the six key volatiles in the MEHSGC (determined for the same sample) is shown in Figure 15. These data are highly correlated. Linear regression of the data gives a line of slope 0.85 and y intercept - 729 wppm. The linear correlation coefficient for this fit of the data is 0.996. Both the good correlation and the slope observed lend confidence to all our other findings that quantities of volatiles sorbed are not at all related to polymer crystallinity. Furthermore, this plot better defines the performance of polymers in this sorption experiment. The extent of sorption, or flavor scalping, increases in the following order:

MDPE < LLDPE ~VLDPE < HDPE < EAA < EVA < IOTEK 4000 < EX906.

The oxygenated polymers clearly perform much worse than the polyethylenes. Among the polyethylenes, the commericially used MDPE performs best. It looks like one toothpaste tube manufacturer may have done some work of their own in this area.

The excellent agreement observed was satisfying but there still remained the issue of why the MEHSGC "total volatiles" data at long contact times are in some cases less than the same data for the shorter contact times. In addition, while other polymers gave more "acceptable" total volatiles at long contact times, i.e. the total volatiles remained constant or increased relative to short contact times, concentrations of specific small molecules such as d-limonene and 1,8 cineole did decrease. The measured total and individual weight losses could be due to migration of molecules from the polymer back into the paste or due to loss from the system either by decomposition or by desorbing from portions of the polymer films which are outside the sealed jar (see Figure 3). The relative contributions of these two possibilities was briefly explored by performing approximate mass balance measurements on two systems. Mass balance was approximately checked for the EX906 and MDPE systems by measuring not only the volatiles concentrations in the

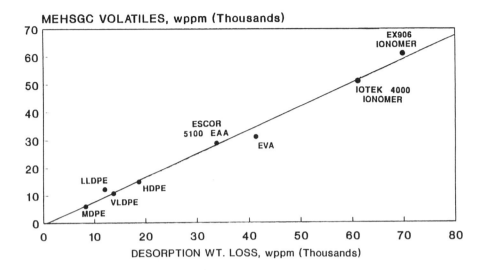

Figure 15.   Gravimetric versus MEHSGC "total" volatiles for weight uptake
of flavor/aroma molecules from toothpaste into polymer films.   These
gravimetric and MEHSGC data are for long contact times, and the gravimetric
data are for measurement of polymer film weight loss (to constant weight) due
to stripping each film of volatiles at 55-70°C under vacuum at the end of the
mint toothpaste contact time.

polymer films but also the volatiles concentrations in the toothpaste at the end of
each experiment. The sum of the (weighted average) concentration from the [film +
toothpaste] at the end of each experiment was then compared with the original
concentration in the toothpaste. The comparisons are shown in Table VI. Both
comparisons indicate that, for the most part, the concentration of each molecule is
nearly constant or declines at most 30% over the course of the sorption experiment.
So while there does indeed appear to be some loss from the system for the smaller,
more volatile molecules, these limited mass balance data indicate that in most
instances all the molecules of a given flavor-aroma molecule are retained in the
system. Consequently, the observed decreases in concentration of some flavor-
aroma molecules in the polymer films at long contact times indicate that flavor-aroma
molecules can migrate back into the toothpaste. We can only speculate at the present
time that this unusual behavior must involve changes in the structure of the polymer
film involved.

Conclusions

In this example of sorption from a complex paste into a polymer, the relationship
found for simple sorbant-polymer systems -- that quantity sorbed is proportional to
percent amorphous polymer -- does not hold. Instead, specific interactions and
solubility parameter matches appear to determine, in part, the outcomes of these
experiments, but in ways we do not yet understand particularly well. Thus, the

Table VI. Mass Balance, Concentrations of Volatiles Aroma/Flavor Molecules in Mint Toothpaste Before Scalping Experiment vs. Sums of Concentrations of Volatiles Aroma/Flavor Molecules -[Toothpaste + EX906] After Long Contact Times

CONCENTRATIONS, wppm[a]

	d-limonene carvone	1,8 cineole cinnamaldehyde	menthone	menthol		
Before (toothpaste)	135	123	400	1513	1397	278
After (toothpaste + EX906)	97	85	365	1708	1319	256
After (toothpaste + MDPE)	171	73	330	1644	1105	279

a   By MEHSGC; reproducibility ± 5-10%

oxygenated polymers sorb 2-8 times more of the volatile flavor and aroma molecules than do the polyethylenes. The order, in increasing quanity sorbed, is:

MDPE < LLDPE ~VLDPE < HDPE < ESCOR 5100 < 18% EVA < IOTEK 4000 < EX 906.

Furthermore, sorption is very inhomogeneous; ratios of quantities scalped are not in simple direct proportion to their ratio in the toothpaste. For example, among the oxygenated polymers, menthol is substantially more extensively scalped than menthone. It also appears that, in these complex systems, scalping can decrease over time from a maximum value. In other words, flavor and aroma molecules can move into the polymer and then later migrate back into the paste. All these findings suggest that the polymer [structure-morphology] changes in response to the ingress of the small organic molecules.

The present study amply illustrates just how complex and unpredictable flavor scalping can be.

Literature Cited

1.   Billingham, N. C.; Calvert, P. D.; Manke, A. S., *J. Appl. Polym. Sci.*, 26, 3543 **(1981)**.
2.   Feldshtein, L.S.; Kuzminskii, A.S., *Vysokomol. Soed.* A13, 2618 (1971).
3.   Kolb, B.; Chromatographia, 15, 587 (1982).
4.   Auer, M.; Kolb, B.; P. Pospisil, *J. Chromatography*, 204, 371 **(1981)**.
5.   Hagmann, A; Jacobsson, J.; Karlson, K., *J. High Res. Chromatog. and Chromatog. Commun.*, 11, 46 **(1988)**.
6.   Hagmann, A.; Jacobson, S., *J. High Res. Chromatog. and Chromatog. Commun.*, 11, 830 **(1988)**.
7.   Barrer, R. M., *Diffusion in Polymers*, Chapter 6, J. Crank and G. S. Park, eds., Academic Press, New York 1968, pp. 165-217.
8.   *Polymer Permeability*, J Comyn, ed., Elsevier Applied Science, New York 1985, p. 65 Solubility is proportional to the amorphous fraction of the polymer.
9.   Kamath, P. M.; Wakefield R. W., *J. Appl. Polm. Sci.*, 9, 3153 **(1965)**.

RECEIVED March 22, 1991

INDEXES

Author Index

Affiliation Index

Subject Index

Production: Margaret J. Brown
Indexing: Deborah H. Steiner
Acquisition: Barbara C. Tansill
Cover design: Amy Hayes

Printed and bound by Maple Press, York, PA

Bestsellers from ACS Books

The ACS Style Guide: A Manual for Authors and Editors
Edited by Janet S. Dodd
264 pp; clothbound, ISBN 0–8412–0917–0; paperback, ISBN 0–8412–0943–X

Chemical Activities and Chemical Activities: Teacher Edition
By Christie L. Borgford and Lee R. Summerlin
330 pp; spiralbound, ISBN 0–8412–1417–4; teacher ed. ISBN 0–8412–1416–6

Chemical Demonstrations: A Sourcebook for Teachers,
Volumes 1 and 2, Second Edition
Volume 1 by Lee R. Summerlin and James L. Ealy, Jr.;
Vol. 1, 198 pp; spiralbound, ISBN 0–8412–1481–6;
Volume 2 by Lee R. Summerlin, Christie L. Borgford, and Julie B. Ealy
Vol. 2, 234 pp; spiralbound, ISBN 0–8412–1535–9

Writing the Laboratory Notebook
By Howard M. Kanare
145 pp; clothbound, ISBN 0–8412–0906–5; paperback, ISBN 0–8412–0933–2

Developing a Chemical Hygiene Plan
By Jay A. Young, Warren K. Kingsley, and George H. Wahl, Jr.
paperback, ISBN 0–8412–1876–5

Introduction to Microwave Sample Preparation: Theory and Practice
Edited by H. M. Kingston and Lois B. Jassie
263 pp; clothbound, ISBN 0–8412–1450–6

Principles of Environmental Sampling
Edited by Lawrence H. Keith
ACS Professional Reference Book; 458 pp;
clothbound; ISBN 0–8412–1173–6; paperback, ISBN 0–8412–1437–9

Biotechnology and Materials Science: Chemistry for the Future
Edited by Mary L. Good (Jacqueline K. Barton, Associate Editor)
135 pp; clothbound, ISBN 0–8412–1472–7; paperback, ISBN 0–8412–1473–5

Personal Computers for Scientists: A Byte at a Time
By Glenn I. Ouchi
276 pp; clothbound, ISBN 0–8412–1000–4; paperback, ISBN 0–8412–1001–2

Polymers in Aqueous Media: Performance Through Association
Edited by J. Edward Glass
Advances in Chemistry Series 223; 575 pp;
clothbound, ISBN 0–8412–1548–0

For further information and a free catalog of ACS books, contact:
American Chemical Society
Distribution Office, Department 225
1155 16th Street, NW, Washington, DC 20036
Telephone 800–227–5558

Highlights from ACS Books

Good Laboratory Practices: An Agrochemical Perspective
Edited by Willa Y. Garner and Maureen S. Barge
ACS Symposium Series No. 369; 168 pp; clothbound, ISBN 0–8412–1480–8

Silent Spring Revisited
Edited by Gino J. Marco, Robert M. Hollingworth, and William Durham
214 pp; clothbound, ISBN 0–8412–0980–4; paperback, ISBN 0–8412–0981–2

Insecticides of Plant Origin
Edited by J. T. Arnason, B. J. R. Philogène, and Peter Morand
ACS Symposium Series No. 387; 214 pp; clothbound, ISBN 0–8412–1569–3

Chemistry and Crime: From Sherlock Holmes to Today's Courtroom
Edited by Samuel M. Gerber
135 pp; clothbound, ISBN 0–8412–0784–4; paperback, ISBN 0–8412–0785–2

Handbook of Chemical Property Estimation Methods
By Warren J. Lyman, William F. Reehl, and David H. Rosenblatt
960 pp; clothbound, ISBN 0–8412–1761–0

The Beilstein Online Database: Implementation, Content, and Retrieval
Edited by Stephen R. Heller
ACS Symposium Series No. 436; 168 pp; clothbound, ISBN 0–8412–1862–5

Materials for Nonlinear Optics: Chemical Perspectives
Edited by Seth R. Marder, John E. Sohn, and Galen D. Stucky
ACS Symposium Series No. 455; 750 pp; clothbound; ISBN 0–8412–1939–7

Polymer Characterization:
Physical Property, Spectroscopic, and Chromatographic Methods
Edited by Clara D. Craver and Theodore Provder
Advances in Chemistry No. 227; 512 pp; clothbound, ISBN 0–8412–1651–7

From Caveman to Chemist: Circumstances and Achievements
By Hugh W. Salzberg
300 pp; clothbound, ISBN 0–8412–1786–6; paperback, ISBN 0–8412–1787–4

The Green Flame: Surviving Government Secrecy
By Andrew Dequasie
300 pp; clothbound, ISBN 0–8412–1857–9

For further information and a free catalog of ACS books, contact:
American Chemical Society
Distribution Office, Department 225
1155 16th Street, NW, Washington, DC 20036
Telephone 800–227–5558